W9-DEA-885

DATE DUE

			WITHDRAWN
			PRINTED IN U.S.A.

WITHDRAWN

Materials and Methods
for
Contemporary Construction

Materials and Methods
for
Contemporary Construction

CALEB HORNBOSTEL

Adjunct Associate Professor, Temple University
Practicing Architect

WILLIAM J. HORNUNG

Founder and Director
Long Island Technical School

Prentice-Hall, Inc., *Englewood Cliffs, New Jersey*

HARRINGTON INSTITUTE OF INTERIOR DESIGN
410 S. Michigan Avenue
Chicago, Illinois 60605

691.
h813
#521

Library of Congress Cataloging in Publication Data

Hornbostel, Caleb.
    Materials and methods for contemporary construction.

    1. Building materials.   2. Building.
I. Hornung, William J., joint author. II. Title.
TA403.6.H67   691   73-18372
ISBN   0-13-560896-1

© 1974 by Prentice-Hall, Inc., Englewood Cliffs, New Jersey

All rights reserved. No part of this book may be
reproduced in any form or by any means without
permission in writing from the publisher.

Printed in the United States of America

10 9 8 7 6 5 4

Prentice-Hall International, Inc., *London*
Prentice-Hall of Australia, Pty. Ltd., *Sydney*
Prentice-Hall of Canada, Ltd., *Toronto*
Prentice-Hall of India Private Limited, *New Delhi*
Prentice-Hall of Japan, Inc., *Tokyo*

410 S. Mich
Chicago, Ill.

Contents

Preface

This book provides the key to an understanding of modern materials and methods of construction. Every student of architecture, architectural technology, or construction technology should have the basic knowledge and information provided within this text. Building materials and methods of construction, both current and standard, are discussed and studied as are the fundamental characteristics of the materials and the various methods of construction in which these materials are used in the construction industry.

The text is composed of 32 study units requiring approximately two semesters of two-hour periods. Each unit is composed of the following:

"Introduction," which provides the student with an overall picture of the contents of the unit; *"Technical Information,"* which covers the body of the subject matter; *"Review Examination,"* in which the student has the opportunity to test himself and discover how well he has learned and retained the subject material (answers for all Review Examinations are supplied at the back of the book); *"Assignment,"* which provides additional practice and study of the subject matter; and *"Supplementary Information,"* which provides further study and enrichment of the student's understanding of the subject. In addition the text includes not only suggested mid-term examinations for the first eight units and for units 17 through 24, but also suggested end-term examinations for units 9 through 16 and 25 through 32. Experience has shown that such mid-term and end-term review examinations often provide the necessary study stimuli for successful completion of the course (answers are supplied for both the mid-term and end-term examinations at the back of the book).

Materials and Methods for Contemporary Construction is intended as a required course in the study of architecture, architectural technology, and construction technology. It is by no means an all-inclusive text, but rather encompasses the most important materials and methods of construction in present-day use. The development of the text parallels to a large degree the actual construction of buildings, beginning with the building process — property, finance, professional services, contracts, and so forth; through the steps ordinarily taken by the contractor in constructing a building; and on up to the final completion and occupancy of the building.

The authors are indebted to the various manufacturers of the numerous building materials whose suggested methods and applications, installations, and final treatments have greatly contributed to the value of the text. Special consideration and appreciation is extended to Dana L. Harris for his assistance in the preparation of the drawings, to Dorothy M. Larkin for her patience and skill in preparing the manuscript through its many stages to the final copy, and to Barbara Hornbostel for editing and proofreading of the text.

Caleb Hornbostel
William J. Hornung

Materials and Methods
for
Contemporary Construction

1

The Building Process

INTRODUCTION

When an individual, group, company, firm, or corporation envisions the construction of a building or buildings, a process is set in motion which involves property, finance, professional services, consultants, lawyers, contractors, subcontractors, material suppliers, furnishing, decorating, site improvement, landscaping and planting, etc., until the building is ready for occupancy. This unit will cover in broad general terms the entire building process.

TECHNICAL INFORMATION

The Property

The property on which the construction is to take place must be checked for size, easements, lot-block-section numbers, zoning, utilities, and the restrictions set by the town, county, city, state, and federal codes.

The legal description and the survey, including the title guarantee, will give the sizes, the easements, and the lot, block, and section numbers (see Figure 1-1).

The building code and the Building Department will give the zoning and building restrictions. An architectural survey (see Figure 1-2) must be made in order to locate utilities, grades, levels, sidewalks, curbs, etc., and a series of soil tests will have to be performed to obtain the bearing capacity of the soil and subgrade conditions.

The Finance

Depending upon the type of building or buildings to be constructed, a financial feasibility study is usually made. For a family residence, this study is not complicated. The family knows the size of house they want, they can visit some builders' houses to check size and cost, they own the property, and can easily check on how much they can afford to spend for the residence and then set a budget.

A large building or group of buildings requires a complete feasibility study. For example, when a company contemplates building a group of garden apartments on a tract of land, the following considerations must be fully evaluated to

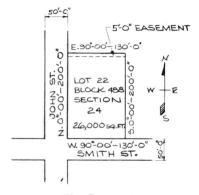

The Property

Starting at the northeast corner of the intersection of John and Smith Streets, thence running north 200'-0" to a point on the east side of John Street, thence east 130'-0" to a point, thence south 200'-0" to a point on the north side of Smith Street, thence west 130'-0" on the north side of Smith Street to the original starting point.

Legal Description

Figure 1-1

3

Information Necessary for a Feasibility Study

Taxes	Town, city, county, state and federal
Cost of Land	Original cost and cost when developed
Building Costs	Sq. ft. and cu. ft. costs for this type of building
Size and Height of Buildings and Types of Apartments	Codes checked for zoning and restrictions
Site Improvement	Grading, excavating, landscaping, planting, roads, walks, etc.
Legal Fees	Lawyers, permits, and contracts
Utilities	Costs for supplying utilities
Insurance	Fire, theft, damage, liability, etc.
Landscaping and Planting	Roads, walks, sidewalks, curbs, parking, grading, seeding and planting
Promotion and Publicity	Advertisement, photographs, renderings, brochures, etc.
Renting and Sales	Staff, office, telephone, etc.
Professional Services	Architects, engineers, decorator, etc.
Maintenance	Costs for maintenance of buildings, painting, repairs
Furnishing and Decoration	Entrance lobbies, halls, stairs, etc.
Profit	Income per year, costs per year
Construction Supervision	Staff for supervision
Mortgages	Construction or investment types: 10, 20, or 30 years
Utilities	Yearly cost of utilities

Professional Services

In some cases the planners and architect or architects will make the feasibility study; in others they will be part of a team, utilized as consultants. When this study is completed, the professional services of planners and architect or architects are obtained in order to produce the drawings and specifications necessary for construction of the building. The general process for producing these drawings and specifications is as follows:

Phase A

1. Schematics, preliminary drawings, specifications, and estimates which include architectural, structural, mechanical, site planning, and landscaping work.

Only after approval of the preliminaries are the next steps taken.

determine the financial feasibility of the project:

2. Architectural working drawings and details.

3. Structural working drawings and details.

4. Heating, ventilating, and air conditioning working drawings and details.

5. Plumbing working drawings and details.

6. Electrical working drawings and details.

7. Site improvement working drawings and details.

8. Landscaping and planting working drawings and details.

9. Separate specifications for each of the above.

10. Final estimate, and final approval before bidding.

Phase B

1. Approval by the Building Department.

2. Approvals by all the other departments and agencies which have specific jurisdiction over building construction.

3. Rendering or model of the project.

4. Preparing forms for competitive bidding. Time of completion, bond, insurance, labor department wage scales, time for bidding, etc.

5. Receiving of bids; consultation and approval.

Phase C

1. Preparing contracts for prime contractors: (a) general construction, (b) heating, ventilating, and air conditioning, (c) plumbing, (d) electrical, (e) site improvement, landscaping, and planting.

In many cases, during the preparation of the working drawings, an interior decorator will be engaged to decorate and furnish the public lobbies, entrances, and halls and to furnish several apartments. The decorator will work with the architect or, as in some cases, be employed by the architect. The furniture, carpets, wall hangings, curtains, etc., will be separately handled by the decorator under separate contracts.

In this entire professional service operation, the coordination between architectural, structural, and mechanical drawings requires very close checking and cross-checking to see that all these various components of a building do not interfere with one another. In Figure 1-3, the space between the hung ceiling and the underside of the concrete floor slab must be of sufficient depth so that duct work can pass under structural beams and girders. In like manner the spacing between ducts and recessed lighting fixtures, as well as plumbing and heating pipes and electrical conduits, must be considered so as not to interfere with their installation.

In the example chosen, a garden apartment project, it is immediately obvious that area limits have to be set on the drawings to designate the limits of property areas that are for the construction of the buildings and under the control of the general contractor, and to mark off at what point he has no further responsibility. Then the site improvement, landscaping, and planting contractor takes over his responsibility of seeding, grading, planting, etc.

Construction

Once costs have been agreed upon and contracts have been signed, the construction operation commences. Because the contracts prepared by the architect give a completion date, it is immediately necessary for the contractor to submit a building construction production schedule in order that all the various building trades know when and at what stage the various parts of their work must be started or completed.

The architect must constantly supervise the construction to check materials, workmanship, and progress in order to issue certificates of payments for the various contractors; the owners usually employ a Clerk-of-the-Works to check these things on a daily basis.

During construction, the inspectors from the various departments having jurisdiction over construction constantly supervise the construction, and in many cases, certain types of work cannot start until approved by them. For example, a private sewage disposal system has to be approved by the Department of Health before work can commence and it cannot be covered with earth until approved by the field inspector.

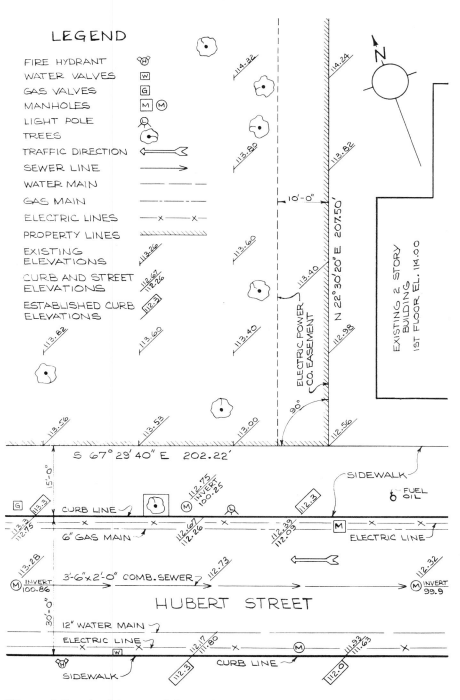

Figure 1-2. Architectural Survey

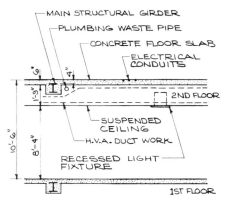

Figure 1-3. Coordination of Architectural, Structural, and Mechanical Components

The various subcontractors and suppliers to the prime contractors will submit samples of materials and shop drawings (fabrication drawings) to the architect for his approval. These drawings have to be checked against the contract working drawings and specifications to see that they meet the standards, materials, etc.

During the construction phase, job meetings are periodically called at which the architect's representative, the Clerk-of-the-Works, and the superintendents of the various prime contractors are present in order to iron out any problems and to check the progress of the job with the construction procedure schedule. At these meetings, the question of changes (change orders) due to job conditions are discussed and decided. For example, suppose that, during the excavating, it was necessary to dig deeper to arrive at the correct bearing soil and therefore more foundation wall, stepped footings, and labor than shown in the contract drawings were necessary. Credits are also discussed. For example, contract specifications may call for a specific type of brick at current cost of so much per thousand; however, the architect and owner select a brick that costs $10 less per thousand.

When the construction phase nears completion, the architect and the Clerk-of-the-Works make up lists of work still to be completed (known as *punch lists*). Here is an example of such an item in a punch list:

Room 208. Door closer missing, piece of base missing on North wall, light fixture frames not installed, sink faucets missing, window glass broken, and switch plates missing.

When all the items on the punch lists are completed and the building is ready for occupancy, a Certificate of Occupancy must be obtained from the Building Department. In general, there are always small items that have to be installed, checked, and replaced after the building is occupied.

As the job reaches completion, there is a final accounting of all extras and credits, and all certificates, guarantees, approvals, etc., must be turned over to the owners. For one year after occupancy, all contractors have guaranteed their work and are responsible for correcting any failure, malfunction, etc.

REVIEW EXAMINATION

1. To obtain the exact size of a piece of property, what should be checked?

2. The legal description and survey including title guarantee give what information concerning a piece of property?

3. In order to locate utilities and grades of a piece of property, what document must be obtained?

4. Name five items that should be considered when making a feasibility study.

5. What do preliminary drawings, specifications, and estimates include?

6. Are separate specifications prepared for architectural, heating, ventilating and air conditioning, and site improvement?

7. Are separate contracts made for general construction and plumbing?

8. Very close coordination between what things is necessary in preparing working drawings?

9. Who supervises the construction of a building?

10. Material samples must be approved by whom?

11. Shop drawings must be checked against what things?

12. When the building is nearing completion, what must the architect prepare?

1. Make a diagram showing the four main parts of the building process and show the connection of the main documents, studies, professional services, and prime contractors.

2. Draw a diagram of a piece of property and write a legal description.

SUPPLEMENTARY INFORMATION

Architectural surveys use two systems for showing grades and elevations, either by contours or spot points, as shown in Figure 1-4.

Architectural professional service fees are generally based on a percentage of the cost of the building construction. The American Insti-

tute of Architects has prepared printed contract forms for owner and architect. They also have available printed contract forms of various types for owner and contractor such as lump sum, cost plus, fixed fee, etc.

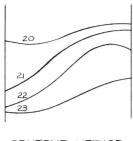

CONTOUR METHOD

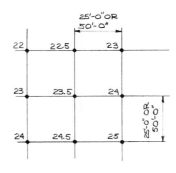

SPOT POINT METHOD

Figure 1-4.   Two Methods of Showing Grades and Elevations

2

Earth Formations

INTRODUCTION

This unit deals with the various kinds of soils that underlie building foundations. In order to calculate the proper size of footing or foundation for a building, the bearing capacity of the soil on which the building is to be placed must be known. But first, the soil composition of the site must be determined.

The soils considered here are: rock, decayed rock, loose rock, shale, slate, boulders, hardpan, gravel, sand, both coarse and fine, clay, mud, silt, and quicksand.

TECHNICAL INFORMATION

Soil

The word "soil" refers to a particular kind of earth or ground. It may be described as the unconsolidated material formed by the disintegration of rock and organic substances in varying proportions. This disintegration of rock masses is *chemical*, due to reaction and decomposition, and *mechanical*, due to frost, moving water, wind, ice, and movement of the earth's crust. The following definitions of the more commonly encountered soils are used in field identification. (Figure 2-1 illustrates these soils graphically.)

Rock. Undisturbed, naturally formed rock masses which are part of the original rock formation. As used in building construction, the term identifies the solid material forming the earth's crust and is

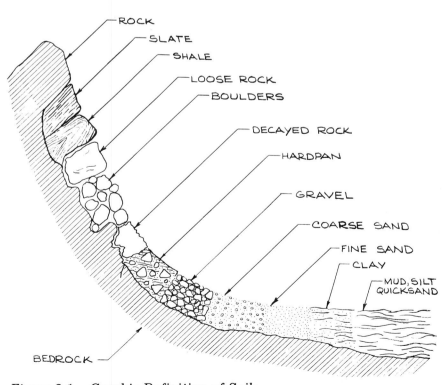

Figure 2-1. Graphic Definition of Soils

9

often referred to as "bedrock" or "ledge rock." Sound hard rock is the best known support for the foundations of a structure.

Slate. A dense, very fine-textured, soft rock which is readily split along cleavage planes into thin sheets and which cannot be reduced to a plastic condition by moderate grinding and mixing with water.

Shale. A laminated, fine-textured, soft rock, composed of consolidated clay or silt, which cannot be molded without the addition of water but which can be reduced to a plastic condition by moderate grinding and mixing with water.

Loose rock. Rock masses detached from the ledge of which they originally formed a part.

Boulders. Detached rock larger than gravel, generally rounded and worn as a result of having been transported by water or ice a considerable distance from the original ledge.

Decayed rock. Sand, clay, and other materials resulting from the disintegration of rock masses, lacking the coherent qualities but occupying the space formerly occupied by the original rock.

Hardpan. A coherent mixture of clay with sand, gravel, and boulders, or a cemented combination of clay, sand, and gravel. "Hardpan," as the word is used in foundation engineering, suggests a degree of hardness caused by consolidation or cementation rather than any definite soil composition. Its removal often requires the use of tools operated by compressed air or even explosives.

Gravel. Detached rock particles 1/4 in. to 3 in. in size, rounded and waterworn.

Sand. Noncoherent rock particles smaller than 1/4 in. Sand has no plasticity or cohesion except along the seacoast and sometimes in beds of streams where sand, gravel, and boulders are generally mixed with other soils. In fact, sand is often referred to as a cohesionless soil. Coarse sand and gravel often furnish an excellent support for building foundations; fine sand is less reliable.

Clay. A fine-grained, firm material resulting from the decomposition and hydration of certain kinds of rock. These finely divided minerals have a sheetlike crystal structure and a chemical composition which give clay its characteristics of slippage and plasticity. In other words, clay is malleable, plastic when wet, and relatively hard when dry. Clay also has elasticity—that is, the ability to regain its original form and dimensions when an applied load that has caused deformation is released. Clay, compared to sand, is relatively impermeable to water.

Mud. Sand, clay, and vegetable and animal matter with water, forming a soft, plastic, and sticky material.

Silt. A finely divided sedimentary material deposited from running water. It consists of sand, clay, and vegetable matter.

Quicksand. Sand mixed with water, forming a mass which yields easily to pressure and sucks down any object resting on its surface. Any uncemented sand, when subjected to the action of moving water, will move; thus any sand moving as the result of the action of water becomes a quicksand.

Effect of Soil on Size of Footings

Table 2-1 summarizes the allowable bearing capacities of various foundation beds. On the basis of these figures, we can now examine the relation between type of soil and size of footings.

The purpose of footings of buildings is to distribute the concentrated loads from columns, piers, or walls over a sufficient area of the soil so that they will not cause

Table 2-1. Allowable Bearing Capacities of Various Foundation Beds, in Tons Per Sq. Ft.*

Hard rock	80
Loose rock	20
Hardpan or hard shale	10
Gravel and sand, well cemented	8
Gravel	6
Coarse dry sand	4
Hard clay	4
Fine dry sand	3
Sand and clay, mixed	2
Wet sand	2
Firm clay	2
Soft clay	1

*Figures given here should be considered as somewhat arbitrary maximums and should be checked by subsurface investigations in actual practice.

objectionable settlement. All foundations, except those on solid rock, will settle to some extent. Slight settlement is of no consequence provided the settlement is uniform, but uneven settlement of foundations may cause serious cracks in floors and walls.

Footings for Lightweight Buildings

Footings under foundation walls distribute the weight of the walls and buildings over a larger ground area, as shown in Figure 2-2. In lightweight construction, the thickness of the footing should be not less than 12 in. and its projection on each side of the wall should be at least 6 in. and not more than one-half the thickness of the footing, in order to avoid undue bending moments and shearing or punching stresses. Footings may extend more than one-half the thickness of the footing when a larger ground area is required to support the loads, provided the footings have reinforcing.

The formula for finding the required number of square feet or area of the footing is *load divided by soil-bearing capacity.* Suppose a load of 40 tons per square foot were imposed on hardpan that has a bearing capacity of 10 tons per square foot—the size of footing would be calculated thus:

$$\frac{load}{soil\text{-}bearing\ capacity} = \frac{40}{10} = 4\ sq.\ ft.$$

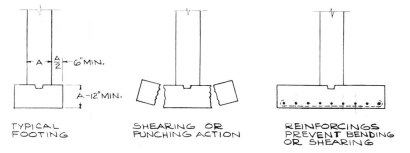

Figure 2-2. Footings for Foundation Walls

Example. Calculate the size of a plain concrete footing (Figure 2-3) under a masonry wall 2'-0" thick, bearing a load of 16,000 lbs. per lin. ft. The soil is firm clay that has a bearing value of 2 tons, or 4,000 lbs. per sq. ft.

$$\frac{load}{soil\text{-}bearing\ capacity} = \frac{16,000}{4,000} = 4\ sq.\ ft.$$

The footing must, therefore, be 4'-0" wide, giving a projection of 1'-0" on either side of the 2'-0" wall. Because the projection should not exceed one-half the depth of the footing, the footing will be 4'-0" wide and 2'-0" deep.

Slab and Pier Footings

When girders are supported on a wall, it often becomes necessary to thicken the wall and footing in order to support the concentrated load. The thickening forms what is known as a pier, as shown in Figure 2-4.

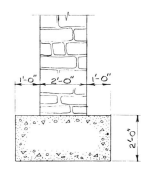

Figure 2-3. Footing for Stone Walls

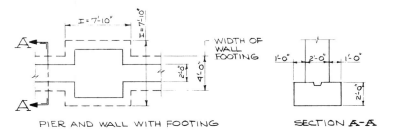

Figure 2-4. Footing for Pier and Wall

The additional areas required on each side of the wall together with the footing area for the wall lying between them should be arranged as a square around the center of

11

HARRINGTON INSTITUTE OF INTERIOR DESIGN
410 S Michigan Avenue
Chicago, Illinois 60605

gravity of the pier. The side of the square may be found from the following formula:

$$I = \frac{b}{2} + \sqrt{A + \left(\frac{b}{2}\right)^2}$$

in which

      I = side of square
      b = width of wall footing
      A = required area of pier footing

Example. See Figure 2-4. A wall $2'-0''$ thick has a footing $4'-0''$ wide. A pier in the wall carries a concentrated load of 240,000 lbs. per sq. ft. The allowable bearing capacity of the soil is 8,000 PSF (pounds per square foot). Find the size of the square footing for the combined pier and wall.

$$A = \frac{240,000}{8,000} = 30 \text{ sq. ft.}$$

$$I = \frac{b}{2} + \sqrt{A + \left(\frac{b}{2}\right)^2}$$

$$I = \frac{4}{2} + \sqrt{30 + \left(\frac{4}{2}\right)^2}$$

$$I = 7.83$$

The length of the side of the combined square footing for wall and pier is 7.83 ft. or about $7'-10''$.

Frost Line

The National Building Code requires that footings be placed at least $1'-0''$ below the frost line.

Table 2-2. *Footing Depth Required by Building Code*

Location	Below Grade	Location	Below Grade
Atlanta, Ga.	$1'-0''$	Jacksonville, Fla.	$1'-0''$
Baltimore, Md.	$3'-0''$	New York, N.Y.	$4'-0''$
Boston, Mass.	$4'-0''$	Omaha, Neb.	$3'-0''$
Chicago, Ill.	$4'-0''$	St. Paul, Minn.	$4'-0''$

Frost line levels vary in different parts of the country, from 0 ft. to 6 ft. deep. In Chicago, the required footing depth is $4'-0''$ because the frost line can penetrate to a depth of $3'-0''$.

Table 2-2 shows the required depths of footings for different areas throughout the United States.

REVIEW EXAMINATION

1. Undisturbed rock masses forming part of the original rock formation is the definition of what?

2. What is the definition of loose rock?

3. Detached rock particles 1/4 to 3 in., rounded and waterworn, is the definition of what?

4. Sand is defined as rock particles smaller than what size?

5. What is defined as sand clays and other materials resulting from the disintegration of rock masses?

6. A coherent mixture of clay with sand, gravel, and boulders is the definition of what?

7. One square foot of solid sound rock supports how many tons?

8. The bearing value of hardpan is how many tons?

9. The soil that supports 4 tons is what?

10. Name the soils that support from 10 to 80 tons per square foot.

11. What is the definition of gravel?

ASSIGNMENT

1. Draw a sketch showing how reinforcing rods are used in footings.

2. Draw a sketch showing a pier in a foundation wall.

3. Prepare a sketch symbolizing the following: (a) hardpan, (b) gravel, (c) coarse wet sand, (d) soft clay.

In addition to the soils defined in this unit, other soils are often mentioned in the building field. These are as follows:

Dirt. Word loosely used to describe any earthy material.

Earth. Material capable of supporting vegetable life and generally limited to material containing decayed vegetable or animal matter.

Loam. Earthy material containing a large proportion of vegetable matter.

Peat. Compressed and partially carbonized vegetable matter.

Soil. Earthy material suitable for growing plant life.

Humus. Earthy material containing a large proportion of decayed vegetable matter.

Topsoil. Earthy material capable of supporting vegetable life and generally limited to material containing vegetable or animal matter in large quantities.

Subsoil. Earthy material not capable of supporting vegetable life.

Alluvial Soil. Sand and clay deposited by flowing water, especially along the river bed.

3

Soil Tests

INTRODUCTION

Before footings for buildings can be designed, the bearing capacity of the soil on which the footings are to rest must be determined. Soil investigation will disclose the types of soil, the water table, the thickness of the bearing stratum, and the uniformity of soil deposits. Subsurface conditions and bearing capacities can be examined by various methods such as

1. Test pits
2. Loading platform
3. Sounding rod
4. Auger borings
5. Wash borings
6. Dry sample borings
7. Rock drilling

This unit deals mainly with the loading platform and the test pit as means of determining bearing capacity and type of soil. It also gives general information on test borings.

TECHNICAL INFORMATION

The type of soil test used depends on general information concerning water table, geographical location and topography, but especially on the size of the building and the loads, as well as information obtained about the conditions encountered when adjoining buildings were constructed.

Of all the methods listed, the test pit and the loading platform give immediate and sufficient answers to permit one to begin designing the footings for the building. All the other methods can be categorized as "test borings" in general terms. These are more complicated methods for unknown terrain and are described in detail, along with the required log of test borings, under the Supplementary Information section.

Test Pits

This method (Figure 3-1) is one of the simplest and probably the

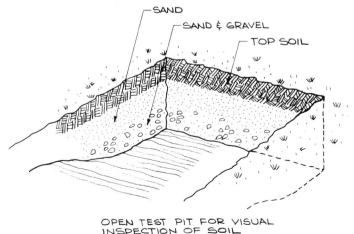

OPEN TEST PIT FOR VISUAL
INSPECTION OF SOIL

Figure 3-1.   Open Test Pit

15

best for determining the nature of the soil because it makes possible the visual inspection of the soil layers, the actual conditions that will be met in building operation, and the compactness and the water content of the soil. It also provides access to the pit for inspection and for obtaining undisturbed samples for lab investigation.

Loading Platform

Loading platform tests are made to assist in determining the soil's safe bearing capacity. When it is not known to what extent the supporting power of a given soil varies with the area subjected to the unit load, and tests on small areas (test pits) are not a reliable guide for the safe load on large areas, loading platform tests are performed.

The loading platform (Figure 3-2) consists of a vertical timber or post which carries a platform to receive the test load, and which has four horizontal guys at the top to keep the post in a vertical position. The bottom of the post forming the loading area should be 24 in. × 24 in., making an area of 4 sq. ft. The platform should be concentric with the post and as close to the bottom of the post as practicable; it may be loaded with pig iron, cement or sand in bags, or any other convenient material.

Testing procedure. Apply sufficient load uniformly on platform as shown in Figure 3-3 to produce a center load of four times the proposed design load per square foot.

$$\text{center load} = \text{load on platform} \times \frac{B}{A+B}$$

Read settlement every 24 hrs. until no settlement occurs in 24 hrs. Add 50% more load and read settlement every 24 hrs. until no settlement occurs in 24 hrs. Settlement under the proposed load should not show more than 3/4 in. Increment of settlement under 50% overload should not exceed 60% of settlement under proposed load. This is shown in Figure 3-4. If the above limitations are not met, repeat test with reduced load.

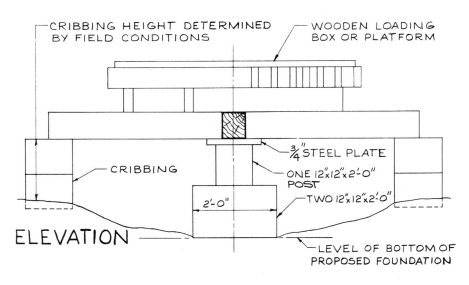

Figure 3-2. Loading Platform

Sample problems

Problem 1: Assume a 12,000-lb. (6-ton) load on platform

$$A = 5 \text{ ft.}$$
$$B = 8 \text{ ft.} \text{ (see Figure 3-2)}$$

Find center load:

$$\text{center load} = \text{load on platform} \times \left(\frac{B}{A + B}\right)$$

$$\text{center load} = 12,000 \times \left(\frac{8}{5 + 8}\right)$$
$$= 12,000 \times 0.615$$

Simplify by using 0.6, thus

$$\text{center load} = 12,000 \times 0.6$$
$$= 7,200 \text{ lbs.}$$

If 7,200 lbs. are carried by 4 sq. ft., logically then 1 sq. ft. carries 7,200/4 = 1,800 lbs. This 1,800 lbs. is the design load of the structure per sq. ft., which means that the bearing capacity of soil tested is 1,800 lbs. per sq. ft.

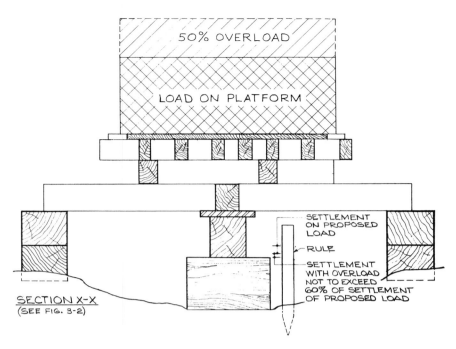

Figure 3-3. Overload Added to Platform

Problem 2: Assume a soil-bearing test is to be made for a design load of 36,000 lbs.: find (a) the center load, (b) the load on platform, and (c) the dimensions of the platform.

(a) Find the center load:

Assume A = 5 ft., B = 8 ft.

$$\text{center load} = 4 \times \text{design load}$$
$$\text{center load} = 4 \times 36,000 = 144,000 \text{ lbs.}$$

(b) Find the load on platform:

$$\text{load on platform} = \frac{\text{center load}}{\dfrac{B}{A + B}}$$

$$\text{load on platform} = \frac{144,000}{0.6}$$
$$= 240,000 \text{ lbs.}$$

(c) Find dimensions of platform:

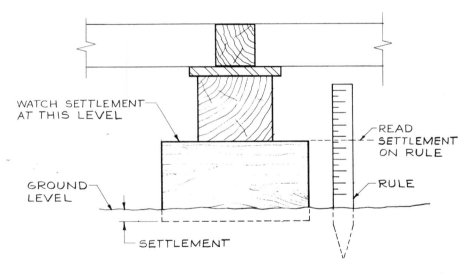

Figure 3-4. Settlement with Overload

Assume the platform is to be loaded with pig-iron blocks, 1 cu. ft., each weighing approximately 477 lbs. The total number of blocks on the platform will be:

$$\frac{240,000}{477} = 503 \text{ blocks}$$

For safe loading, the blocks on the platform should not be stacked higher than the length of the platform, including the overload.

Example. Assume a platform of 15 ft. × 15 ft. The area is 225 sq. ft., which will take 225 blocks in one layer on the platform. Two layers will take 450 blocks, with a third partial layer of 53 blocks, making a total of 503 blocks, 3 ft. high. With a 50% overload, the blocks will not exceed a stacked height of 15 ft. This size platform can be satisfactory.

17

If a platform of 10 ft. × 10 ft. were considered, each layer would take 100 blocks. Total height would be 6 ft. A 50% overload would require another two layers, the total height at this point not exceeding a height of 10 ft., the dimension of the platform. The platform is, therefore, also considered satisfactory.

Log of Test Borings

Most urban areas require test borings as part of the building code requirements. When preliminary designs of the building have been finalized, the architect, through his structural engineers, will make a diagram showing where test borings are to be made (see Figure 3-5). The test boring report, known as the log of test borings (see Figures 3-6 and 3-7), gives the architect and structural engineer the necessary information to design footings, piers, foundations, etc., for the building and also to determine the depth of cellars or basements and whether the entire foundation has to be waterproofed or only damp-proofed.

Test boring reports should give complete information regarding all holes that were made, including also those that were started and then abandoned. The report may be given in the form of a log, showing the location of the holes that were made on the plot plan, the starting elevation of each hole or boring, the depths at which the various strata of soils and rock are encountered, the depth of ground water, and the presence of fill or other materials. The borings should be made to a reasonable depth below the contemplated footings so that the true nature of the soil on which the footings are to rest is known.

Soil-boring logs are often included on the contract drawings so that the contractor cannot ever claim that information was withheld from him.

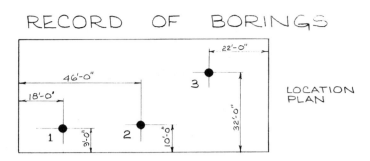

RECORD OF BORINGS

Figure 3-5. Plot Plan Showing Location of Test Borings

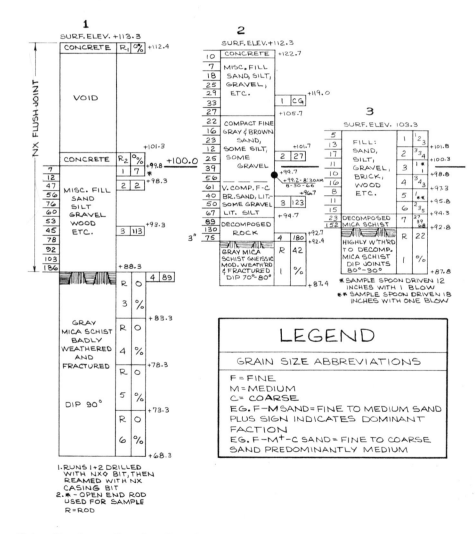

Note: For Spoon Samples, Core Drilling and Equipment, see Fig. 3-7.

Figure 3-6. Log of Test Borings

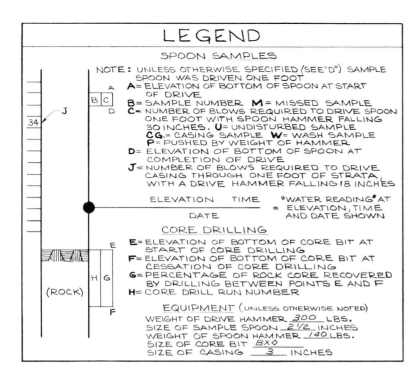

Figure 3-7. Log of Test Borings (cont.)

REVIEW EXAMINATION

1. Name at least three methods by which soil can be tested for its bearing capacity.

2. What are cribbing blocks used in connection with?

3. The center load is equal to what?

4. The design load is the load of what in sq. ft.?

5. What is the center load when the load on the platform is 12,000 lbs.? (Distance A = 5 ft.; distance B = 8 ft.)

6. In the above problem, one cribbing block carries how many pounds?

7. If the load on the platform is 12,000 lbs., what is the load on the platform with the overload?

8. Apply a sufficient load on the platform to produce a center load of how many times the proposed design load; add 50% overload and read settlement every how many hours until no settlement occurs in how many hours.

9. Settlement under load on platform should not show more than a settlement of how many inches?

10. Settlement under 50% overload should not exceed what percent of settlement under proposed load?

11. If the above limitations are not met, the test should be repeated with what load?

12. When test borings are made, what is submitted to the architect?

13. If the soil is to be tested to safely withstand a design load of 10 tons, what load should be placed on the platform? (Assume A = 5 ft., B = 8 ft.)

1. Prepare a sketch, plan, and elevation of the loading box or platform.

2. Make a sketch of a building on a piece of property and locate by dimension where you want test borings to be made.

SUPPLEMENTARY INFORMATION

Sounding Rod

The sounding rod test is used not to bring up soil samples, but only to determine the elevation of sound rock and the approximate length of piles that may be required. A sounding rod is usually in the form of a solid steel rod or pipe, the pipe being driven in lengths of about 5 ft. and provided with a point at the lower end. Driving is often done by hand, occasionally with a hammer operated by a gasoline engine.

Provided that boulders or other obstacles are not mistaken for ledge rock, sounding rods may have a useful but limited application for the purpose of determining the depth to rock or other bearing, such as gravel. The results of such tests, however, should not be relied upon except in special cases where the character of the soil is already known and the only object of the investigation is to determine the depth to a firm stratum at various points upon the site. For example, sounding rods are used where land has been filled and only the depth to the firm bearing soil has to be determined.

Auger Boring

The auger boring test is designed to bring up soil samples from shallow foundations in sand and clay, or any other soil that will stick to the bit of the auger, for removal. Auger borings are made with a 2 or 2-1/2-in. auger fastened to a long pipe which is encased in a larger pipe. After a number of turns, the auger is removed, bringing up samples of the soil.

Such borings give a fairly good description of the types of soil, such as rock, boulders, or hardpan containing rock. Auger borings are most practical when used in sand or hard clay, but are difficult to use in coarse gravel and soft clay; they are not practical for depths exceeding 18 to 24 ft.

Wash Boring

This method is used to bring up both dry and wet samples. It is not as efficient as other methods, because the soil that is brought up is thoroughly mixed, one layer with another. It is useful, however, in locating bedrock and when the material is too compact for good results with an auger.

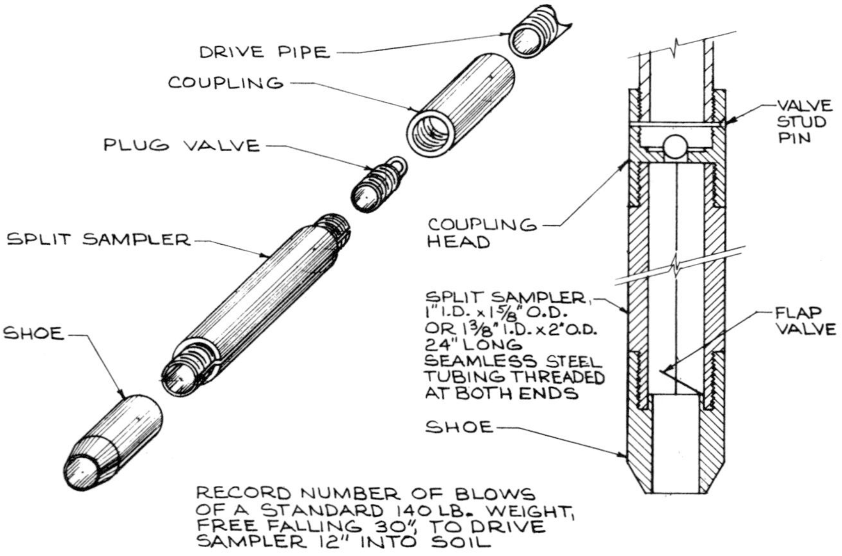

DRIVE PIPE
COUPLING
PLUG VALVE
SPLIT SAMPLER
SHOE

VALVE STUD PIN
COUPLING HEAD
SPLIT SAMPLER, 1" I.D. x 1⅝" O.D. OR 1⅜" I.D. x 2" O.D. 24" LONG SEAMLESS STEEL TUBING THREADED AT BOTH ENDS
SHOE
FLAP VALVE

RECORD NUMBER OF BLOWS OF A STANDARD 140 LB. WEIGHT, FREE FALLING 30", TO DRIVE SAMPLER 12" INTO SOIL

Figure 3-8. Dry Sample Boring

Wash borings are made with a pipe, 2 to 4 in. in diameter, driven into the soil and containing a smaller jet pipe through which water is forced. The flow of water washes the material at the bottom up to the surface, where it is collected and tabulated. The finer materials, such as clay, sometimes disappear in the washing and the heavier materials separate from each other, thus reducing the dependability of the samples. Wash borings may also be stopped by boulders which are then sometimes mistaken for bedrock. Wash borings can penetrate all other materials, however, and can be carried downward 100 ft. or more, and are often sufficiently reliable.

Dry Sample Boring

This method employs a pipe forced down into the soil. A smaller pipe, which is inserted into the larger pipe, has a sampling spoon at the bottom instead of a drill. The spoon is driven down and the pipe is lifted for the sample (see Figure 3-8). The contents of the 2-in. pipe are removed and placed in a jar or bag for laboratory testing. Samples of soil are usually taken every 5 in. by the dry boring method. Boring tests must show the nature of the soil in at least one location for every 2,500 sq. ft. of building area (see Figure 3-5).

Rock Drilling

This method is necessary when rock is encountered in order to make certain that the boulder or thin layer of rock is not mistaken for bedrock. This means that the rock must be drilled for some distance until the core sample discloses bedrock. Core samples are placed in flat metal or wood containers (see Figure 3-9) and brought to the testing laboratory.

Core borings, also known as diamond drill borings, are more costly than other methods of testing but are the most dependable. They can penetrate to great depths, through all materials including rock, and bring up complete cores or cylinders of the material through which they pass. Books on soil mechanics or soil engineering may be consulted for a more detailed description of how these tests are performed and the equipment that is used.

Ground Water

Below a certain level all soil is saturated with ground water. This level is often called the water table; it roughly follows surface contours but may be lowered by dry seasons, by pumping, or by covering the surface with buildings and pavements. Where there is any possibility of an actual hydrostatic head occurring at the level of the bottom of the floor slab, provisions must be made to relieve or resist the resulting pressure, and an appropriate type of waterproofing installed (see Figure 3-10).

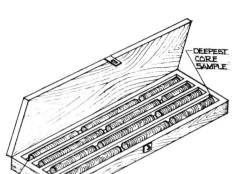

Figure 3-9. Rock Core Samples

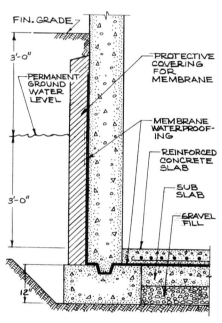

Figure 3-10. Waterproofing Membrane

4

Utilities

INTRODUCTION

This unit deals with the study of the public and private utilities for urban, suburban, and rural areas. Description of the typical plumbing drainage system is included, with definitions of the various parts of the system.

TECHNICAL INFORMATION

The phrase "public utilities" means the services that are supplied to the consumer by private utility companies, by the various departments of the city, town, and community, or, in some areas, by federal agencies for electrical power such as the Tennessee Valley Authority (TVA). Such services include

1. Sewage disposal and storm-water drainage system
2. Supply of potable water
3. Gas for heating and cooking
4. Electricity for lighting, heating, and other uses
5. Telephone service

Public Utilities in the Urban (City) Areas

In Figure 4-1 are shown the typical utilities located within the street in an urban area. The house sewer lines from the various buildings connect to the main sewer within the center of the street. The water main is usually located 9 ft. from the curb and 4 ft. deep, while the gas main is usually 6 ft. from the curb and 3 ft. deep. Electrical conduits are also provided for electricity and telephone lines. There is a separate water system for fire hydrants.

Rainwater on sidewalks and on streets flows along the curb and is conducted into a catch basin which,

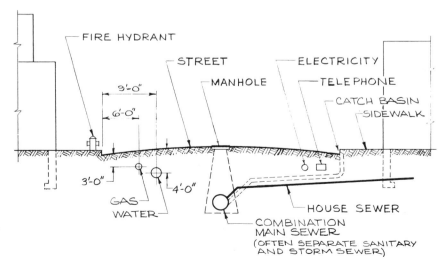

Figure 4-1. Typical Urban (City) Public Utilities

23

in turn, may be connected to the public or main sewer. The main sewer then is known as a combination storm and sanitary sewer system. The storm sewer may, in certain areas, be independent of the sanitary sewer and therefore drains into its own piping system within the street.

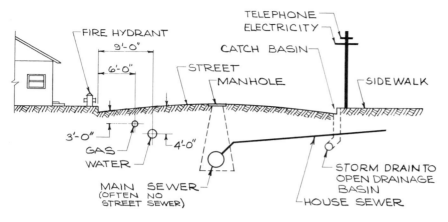

Figure 4-2. *Typical Suburban Public Utilities*

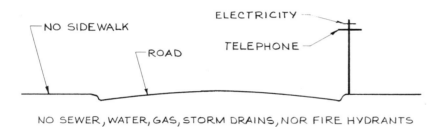

Figure 4-3. *Typical Rural Utilities*

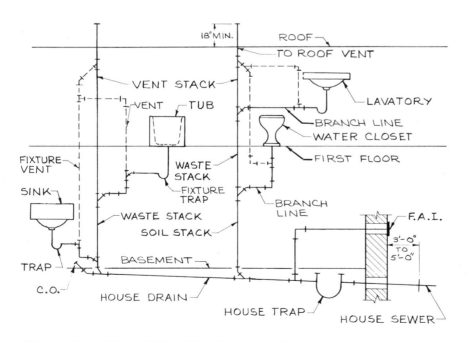

Figure 4-4. *Typical Plumbing Drainage System*

Public Utilities in Suburban Areas

Telephone and electrical lines are carried either by poles above ground as shown in Figure 4-2 or by conduits in the street. Public main sewers may or may not be provided. If no main sewer exists, private sewer systems such as septic tanks with disposal fields or leaching pools are required for each home or other type of building. Gas and water supplies are usually located in the streets. A separate water system is provided for fire hydrants. Rainwater on sidewalks and streets flows along curbs and is conducted into a catch basin which, in turn, feeds into a storm drain which feeds to open drainage basins.

Rural Utilities

Utilities are usually limited to electricity and telephone lines carried by the electric or telephone poles (see Figure 4-3). In the rural home or other type of building, the owner must provide his own sewer system—a septic tank with either one or more leaching pools or a subsoil disposal system. He must obtain his own wellwater, and he uses bottled gas or electricity for cooking and fuel oil for heating.

The Typical Plumbing Drainage System

Figure 4-4 shows a typical plumbing drainage system within the building that conducts the sewage and the waste water to the public sewer or to a private sewage disposal system. This arrangement of the piping is typical for both large and small buildings whether urban, suburban, or rural. The system of pipes, fittings, and fixtures may be listed and defined as follows:

1. house sewer
2. house trap
3. vent for house-trap fresh air inlet (abbreviated as F.A.I.)
4. house drain
5. soil stack
6. waste stack
7. branch line
8. fixture trap
9. fixtures
10. vent lines

House sewer. That portion of pipe leading from the house to the public sewer or private sewage disposal system (see Figure 4-5).

House trap. A device, fitting, or assembly of fittings installed in the building drain to prevent sewage gases from entering and circulating through the building drainage system inside the building (see Figure 4-6).

Fresh air inlet. Required on all plumbing drainage systems where a house trap is used (see Figure 4-6). The fresh air connection is made immediately behind the house trap; its other end is carried to the outside air. Its purpose is to prevent the breakage of the house-trap seal through a syphoning action.

House drain. That part of the lowest piping of the drainage system which receives the discharge of soil, waste, and other drainage pipes inside the walls of the building and conveys such discharges to the house trap and thence to the public sewer or private sewage disposal system (see Figure 4-7).

Soil stack. The vertical pipe that receives the discharges from a water closet.

Waste stack. The vertical pipe that receives only liquid wastes as from a lavatory, tub, shower, or sink.

Branch line. A part of the piping system that connects to a fixture.

Fixture trap. A fitting or device designed and constructed so as to provide, when properly vented, a liquid seal that will prevent the back-passage of air from the drainage system without materially affecting the flow of the sewage or waste water.

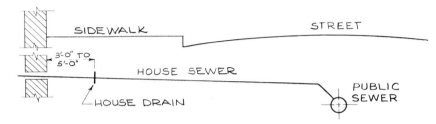

Figure 4-5. The House Sewer

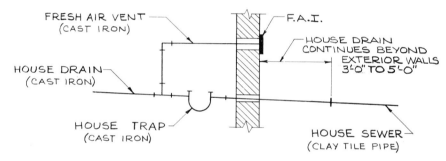

Figure 4-6. The House Trap

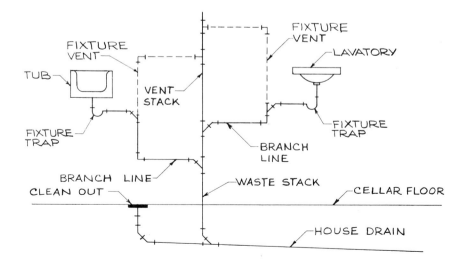

Figure 4-7. The House Drain

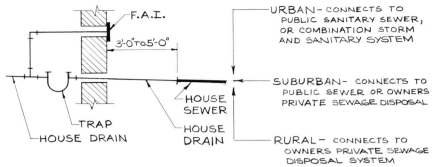

Figure 4-8. *House Sewer Connections*

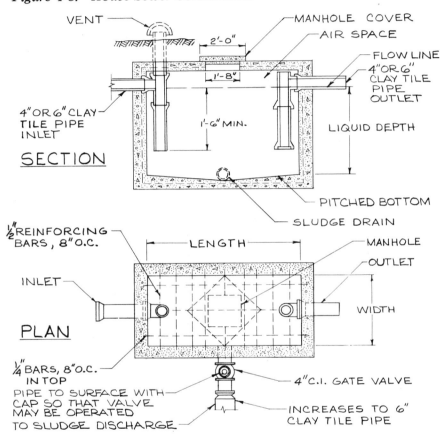

Figure 4-9. *Septic Tank*

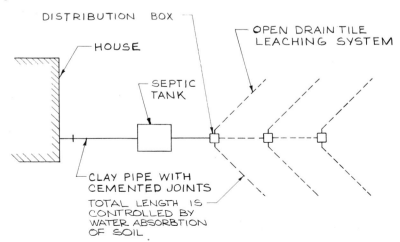

Figure 4-10. *Septic Tank Disposal System*

Fixtures. Lavatories, sinks, water closets, bathtubs, shower stalls, urinals, etc.

Vent lines. That part of the drainage system consisting of piping installed to permit adequate circulation of air in all parts of the sanitary drainage system and to prevent trap syphonage and back pressure.

House sewer connections for urban, suburban, and rural areas may be such as shown in Figure 4-8.

The *septic tank disposal system*, shown in Figures 4-9 and 4-10, does not permit solid matter to enter the soil. Its primary function is to hold the sewage without leakage, to allow thorough digestion of the solid matter, and to retain suspended solids, letting only liquids (effluent) overflow into a leaching pool or a disposal area.

The *disposal area* (see Figure 4-10) consists of field-type pipe, 4 in. in diameter, laid with open joints covered over the metal or tar paper to prevent fill-in by loose earth. The effluent liquid drains from each open joint and is purified by the microbes in the soil. Porous soil makes the best disposal area.

The *leaching pool* may be built with curved concrete block with open joints, as shown in Figure 4-11, or obtained as precast concrete units. The effluent from the septic tank is drained into the leaching pool; there the liquids are absorbed into the soil. The upper portion of the leaching pool is built of corbelled brick or block laid in mortar joints. A cover, 1'-8" in diameter at the top, is placed just below the finished grade.

What Makes A Good Sewage Drainage System?

The principal requirements of a good sewage drainage system may be summarized as follows:

1. It must carry the spent and waste water rapidly away from the fixtures;

2. The passage of air, gas, odors, and vermin from the sewer into the building must be prevented;

3. The drainage pipes must be air-, gas-, and watertight;

4. The system must be so installed that slight movements or vibrations of the building, piping, or fixtures will not cause leakage.

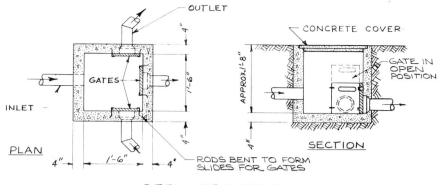

DETAIL OF DISTRIBUTION BOX

DETAIL OF OPEN DRAIN TILE

Figure 4-10 (cont.)

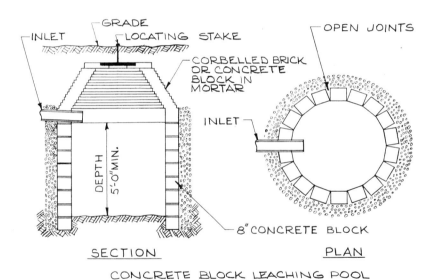

CONCRETE BLOCK LEACHING POOL

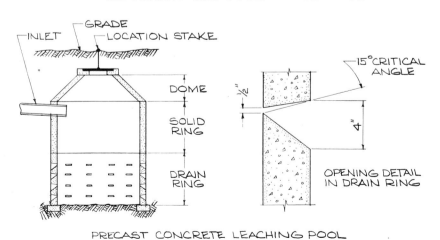

PRECAST CONCRETE LEACHING POOL

Figure 4-11. Leaching Pools

27

COLUMBIA COLLEGE LIBRARY

1. Where are public utilities in an urban area usually located?

2. The water main in a typical urban area may be how many ft. below the street level and how many ft. from the curb?

3. What are three other public utilities?

4. The house sewer is that portion of pipe running from the house drain and connecting to what in the street?

5. In order to prevent the back-up of gases the house trap is always filled with what?

6. The fresh air inlet prevents the water seal in the house trap to be what?

7. What are the typical rural public utilities provided?

8. How does the soil stack differ from the waste stack?

9. The typical rural house plumbing system connects to what?

10. What does the typical suburban house plumbing system connect to?

ASSIGNMENT

1. Prepare a sketch of the typical plumbing system within a building. Label all pipe.

2. Prepare a sketch of a typical street in an urban area with all its utilities.

3. Sketch a typical septic tank with a leaching pool or a disposal field.

SUPPLEMENTARY INFORMATION

Problems of Private Sewage Disposal System

Pollution is a major concern today; private sewage disposal systems must be approved by the health department having jurisdiction in the area where the building is to be constructed. Today, the installation of a septic tank has almost become mandatory, and either leaching pools or subsoil disposal drains are used to get rid of the effluent from septic tanks. The following indicates how a subsoil disposal drain system is calculated (see Tables 4-1 and 4-2).

Test for absorption rate. Dig a test pit $0'$-$12''$ square by $1'$-$6''$ deep within the future subsoil disposal drain area. Quickly fill the pit with 6 in. of water and establish time in seconds until water disappears.

$$\frac{\text{time in sec.}}{6} = \text{sec. required for water to drop 1 in.}$$

Problem

A test pit shows 24 sec. elapsing before water disappears. What is the number of lin. ft. required for the subsoil disposal drainage field for a building having an occupancy of 16 people?

$$\frac{24 \text{ sec.}}{6} = 4 \text{ sec. required for water to drop 1 in.}$$

Table 4-1. Soil Absorption Ratings

	Rapid Absorption	Medium Absorption	Slow Absorption
Seconds required for water to drop 1″	0 to 3	3 to 5	5 to 30

From Table 4-1 we can see that the soil rating is medium absorption, and from Table 4-2 we see that for 16 people the required number of lin. ft. of subsoil disposal drains is 650'-0".

The Fixture Unit

To determine the capacity or size of a drainage and vent pipe plumbing system, a convenient commonly used method is based upon a unit called a *fixture unit*. A fixture unit has been set as a rate of discharge of 7-1/2 gal. per minute from a fixture. This is approximately the rate of discharge from one lavatory. Table 4-3 lists the equivalent fixture units of various fixtures.

Building codes give tables of maximum permissible loads for sanitary drainage piping also in terms of fixture units. For example, the size of piping in any horizontal fixture branch is controlled by the number of fixture units that drain into it. The required pipe diameter is indicated in Table 4-4.

Table 4-2. Length of Subsoil Disposal Drains for Number of People Served*

No. of Persons Served	Linear Feet Required		
	Rapid Absorption	Medium Absorption	Slow Absorption
1–4	100	150	200
5–9	200	350	700
10–14	340	500	1000
15–20	475	650	1250

*Table based on 50 gal. of sewage per person in 24 hrs.

Table 4-3. Equivalent Fixture Units

Fixture	Fixture Units	Fixture	Fixture Units
Lavatory	1	Water closet	6
Kitchen sink	1½	Shower	3
Bathtub	2	Floor drain	3
Urinal	3	Slop sink	4
Laundry tray	3	Roof drainage (180 sq. ft.)	1

Table 4-4. Relation of Pipe Sizes to Fixture Units

No. of Units on Horizontal Branch	Pipe Diameter (in inches)
1	1¼
3	1½
6	2
12	2½
20	3

5

Site Preparation and Excavation

INTRODUCTION

This unit deals primarily with the preparation of the building site. Such preparation entails the removal of old buildings, the installation of the utilities for the new structure, and excavation, including the general excavation for cellars or basements with possible ground water conditions, and the trench excavations required for the utilities. The protection of adjoining structures that may require shoring and underpinning is also considered.

TECHNICAL INFORMATION

The first step in the construction of a building is to prepare the site or property on which the building is to stand. Existing old buildings may need to be demolished and removed from the site, including their footings and foundation walls. Existing sewer, gas, and water lines are cut off and capped, and electric power lines disconnected or, if necessary, entirely removed. Existing trees and shrubs that are to remain on the site should be protected while construction work is in progress.

Figure 5-1 illustrates a plan and section of a site on which a proposed new building, outlined in dotted line, is to be erected. The smaller existing building must be removed because it now occupies the space of the proposed new building. Water, gas, sewer lines and electric power serving the existing building. Water, gas, sewer lines, and much larger pipes and electric power lines will be required for the new building.

Shoring and Underpinning

A part of one wall of the new building, with footings lower than those of the adjacent building, is to be built directly against the adjacent building. Before any general excavation for the new building can be started, it becomes necessary to first lower the footings of the adjacent building to the level of the footings for the proposed new building (see Section A-A, Figure 5-1). This can be accomplished by using temporary supports such as posts, timbers, and beams, to carry the weight of the building until new and lower foundations are in place. The most frequently used types of temporary supports are shores, and

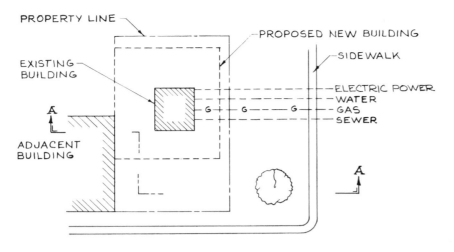

PLAN OF BUILDING SITE

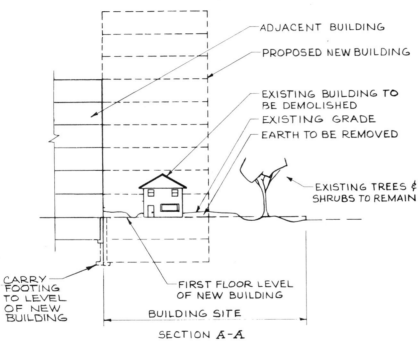

SECTION A-A

Figure 5-1. *Plan of Building Site*

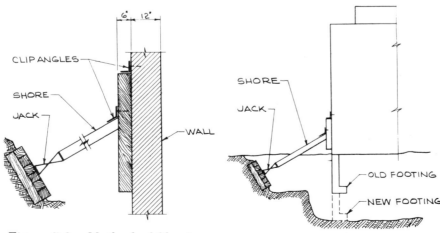

Figure 5-2. *Method of Shoring*

for underpinning, needles or needle beams and pipe cylinders.

Shores consist of long wood posts placed in an inclined position against the wall to be supported (see Figure 5-2). The lower end of the shores rests on a wood platform with a face at right angles to the inclination of the shores. By introducing a jack, the posts or shores can be forced into close bearing against the wall. Shoring thus supports, braces, and stabilizes an existing wall while the construction of the footings and foundations of the new building are installed.

Needles for underpinning are wood or steel beams penetrating holes cut in an existing wall and supported on both sides of the wall on wood blocks or cribbing. Needle beams rest on heavy jacks (see Figure 5-3). The jacks are used to force the needle beams upward, thereby supporting the wall. With such support, the old walls and footings are removed and new footings and walls at their proper levels are installed (see Figure 5-4). When this is done, the weight of the wall is then carried by the new foundation.

Pipe cylinders for underpinning are another method of support used in place of needle beams (see Figure 5-5). Hydraulic jacks, which react against the existing wall, press the sections of pipe cylinders into the ground to refusal and so transfer the weight of the wall to this new underpinning.

The final operation consists of transferring the loads of the old wall onto the new. This operation must be very carefully done in order to avoid any settlement. The usual procedure consists of bringing up the underpinning wall to within 2 or 3 in. of the bottom of the existing wall and filling the remaining space with dry-pack consisting of cement and sand with a minimum of water added. The pack is rammed into place until the gap is filled. The dry-pack has almost no shrinkage, due to the small amount of water. As soon as the entire weight of the wall has been transferred to the new sections of footings and foundations, the shoring and underpinning are removed and are replaced with new footings and foundations.

Excavation

The work of excavation means the removal of all materials of every kind and description for the proper installation of the new foundation. Backfilling around the foundation walls, the spreading of additional fill to the levels indicated on the plans, and the final or finished grading and topping of the site are often included under the work of excavation. Figure 5-6 shows the work of excavation for a typical building. Excavation may be categorized and defined as follows:

Removal of Topsoil. After inspection of the site, if the top layer of the soil is good topsoil, it is removed by scraping the surface to the depth of the topsoil (usually about 6 in.) and depositing this material in a convenient corner of the site. When all backfilling and rough grading are completed, the topsoil is then spread over the area as a final topping for seeding, forming the finish grade. On sites where the topsoil is not worth saving, the builder is required to secure a good topsoil from suppliers of this material.

General Excavation. This includes the excavation for cellars or basements, and floor levels below grade. General excavations may be made with power equipment such as clamshells, hoes, power shovels, or bulldozers. Specifications often state that the contractor is to excavate to the levels indicated on the plans, sections, and details, and to allow sufficient excavation around the foundation for the proper in-

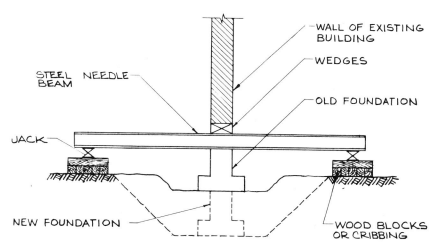

Figure 5-3. The Use of Needle Beams

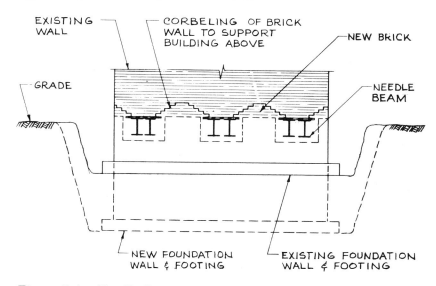

Figure 5-4. Needle Beams Support Wall While New Foundations Are Installed

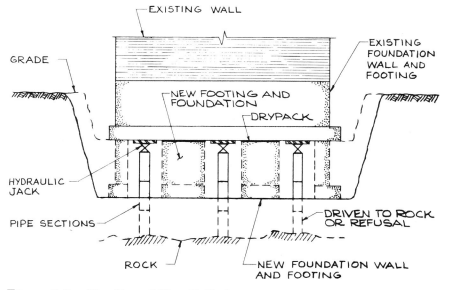

Figure 5-5. The Use of Pipe Cylinders

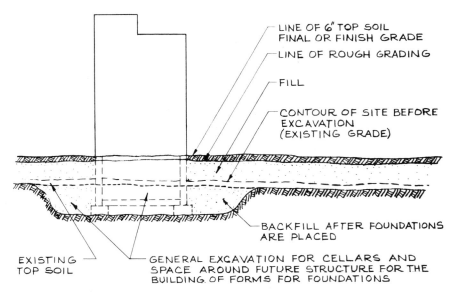

Figure 5-6. Excavation and Backfilling

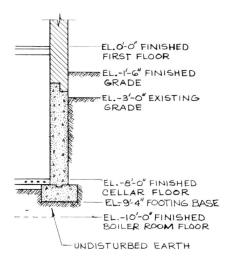

Figure 5-7. Excavation Levels as Shown on Section

those specified for the footings, it becomes the contractor's responsibility to fill in such excess excavation with a suitable layer of concrete (see Figure 5-8), the thickness of which is to be specified by the architect or engineer. Where soil is less cohesive or where space for excavation is limited, bracing is used to support the soil.

Trench excavations. Trenches for footings are generally excavated and trimmed by hand. This trimming is done just prior to pouring of the footings. Trenches for footings and foundations and any trenches required for the installation of sewer, water, gas and electric, and telephone conduits are dug with power equipment to the depths and widths specified.

Where the soil is not firm, trenches may require sheathing and bracing to prevent sliding or cave-in of the soil (see Figure 5-9). Planks may be installed against the excavated soil and braced. Where the soil is less cohesive, sheathing boards or planks are installed and braced on both sides of the walls for the entire length of the trench.

Where concrete columns and piers require deeper levels in order that their footings may be brought to the proper bearing soils, deep pits are excavated which may require lagging boards and bracing such as indicated in Figure 5-10. Because the lagging boards are driven into the ground similarly to piles, the process is called "sheetpiling." Sheet-piling may be of wood or steel, or a combination of both.

Water problems in excavation. All excavations must be kept dry in order to make possible the installation of footings and foundation walls. After heavy rains, water collected in excavations must be pumped out (see Figure 5-11).

Wherever ground water is present and where excavations are carried below ground water and tide levels, it becomes necessary to maintain

stallation of the forms for the foundation walls.

The wall section in Figure 5-7 is typical of the excavation levels as they appear on the plans. The existing grade, or the level of the ground before building is begun, is shown as elevation minus three feet, written $EL - 3'-0''$. The finish or final grade is $EL - 1'-6''$ requiring a fill around the building. The minus elevations for the cellar floor and the finish boiler room floor are shown on $EL - 8'-0''$ and $EL - 10'-0''$ respectively. All minus dimensions are taken from a point marked $EL - 0'-0''$, which often is also the finish first-floor level.

Footings must rest on firm or undisturbed soil that will not settle under the imposed load. When general excavations are inadvertently carried to lower levels than

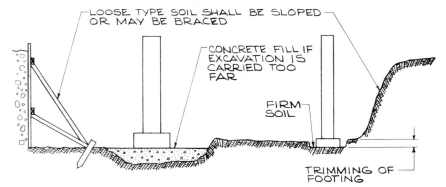

Figure 5-8. General Excavation and Bracing

continuous pumping operations. This is accomplished by the use of well points driven into the ground around the excavation. These well points are piped together above the ground and connected to one or more pumps. By this method the water is pumped out of the ground to lower the ground water level below the depth of the excavation (see Figure 5-12).

Backfilling around foundations. After the forms for the foundation are removed, the area around the walls is backfilled with a soil consisting of sand and gravel, and small rock or earth. It is deposited in layers 12 in. thick, thoroughly compacted to avoid settlement. A well-tamped backfill exerts less pressure against the foundation walls than a loose fill (see Figure 5-13), which may cause failure of foundations, especially during heavy rains when surface water causes loose fill to become liquid and push against the walls.

Excess and additional fill. In preparation for the new building on the site, excess earth must be removed from the site or, when necessary, spread over areas that require fill.

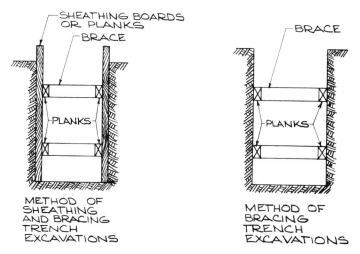

Figure 5-9. *Bracing or Sheathing for Trench Excavations*

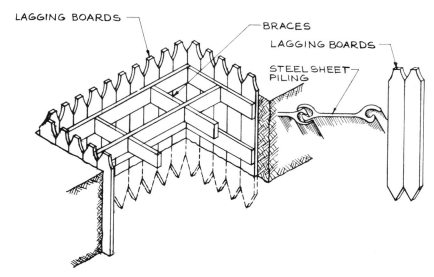

Figure 5-10. *Method of Sheet Piling and Bracing Deep Pits for Columns or Piers*

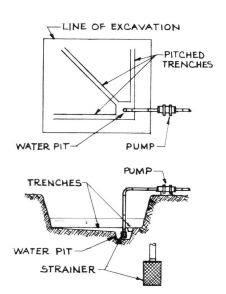

Figure 5-11. Pumping of Water from Excavation

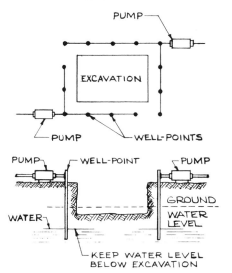

Figure 5-12. The Well-Point System

35

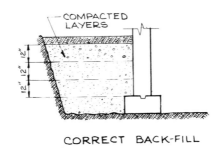

Figure 5-13. Backfill around Walls

REVIEW EXAMINATION

1. When foundations of a building are to be brought to a lower level because of an adjacent excavation, what temporary supports may be used?

2. When an existing building is to be demolished, what is the first operation?

3. What is the first step taken when beginning to excavate a site?

4. The earth excavation from cellars and basements is used for what purpose?

5. The earth fill around foundations should be well-tamped to accomplish what?

6. When excavations are carried lower than called for on the plans, what must be done with the excess excavation under the footings?

7. Where the soil is not self-supporting in trench excavation, what are the trenches required to be?

8. For what purpose are lagging boards used?

9. When is a well point system used?

10. What is the most suitable material for backfilling around a foundation?

ASSIGNMENT

1. Prepare a free-hand sketch illustrating the correct method of backfilling around a foundation wall.

2. Prepare a sketch illustrating a footing and foundation wall with a 4-in. concrete slab or 6-in. gravel fill. Show the finished grade 8 in. above the foundation wall. Locate the minus dimensions of the cellar slab.

3. Make a sketch illustrating bracing and sheet-piling a trench.

SUPPLEMENTARY INFORMATION

Retaining walls. These are independent walls designed to hold back a bank or earth where the grade is sloped. The angle of the slope, known as the angle of repose, is largely controlled by the nature of the soil. Sand, for example, requires a slope angle ratio of 1 to 1-1/2; that is, for every 1-1/2 ft. of horizontal length, the soil rises 1 ft. of height. Other soils require dif-

Table 5-1. Slopes for Various Types of Banked Earth

Type of Earth	Slope of Repose		Angle of Repose
	Length (ft.)	Height (ft.)	
Clean sand	1'-6"	1	33°-41'
Sand and clay	1'-4"	1	36°-53'
Rotten rock	1'-4"	1	36°-64'
Damp soft clay	3'-0"	1	26°-34'

ferent slopes, as indicated in Table 5-1.

The retaining wall in Figure 5-14 is provided with weep holes, 4 to 6 ft. apart, to drain off the water accumulated in the soil and thereby reduce the pressure against the retaining wall.

Banked earth slopes may also be contained by piling up stones or concrete blocks, as shown in Figure 5-15.

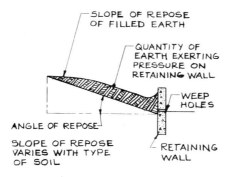

Figure 5-14. The Angle and Slope of Repose

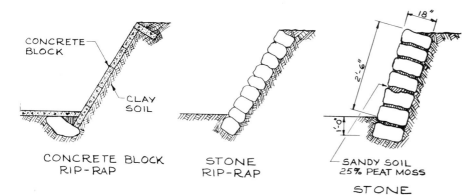

Figure 5-15. Banked Earth Slopes Supported by Concrete Blocks or Stone

6

Footings and Foundations

INTRODUCTION

The weight of a building is transferred through foundations, piers, and columns onto footings which must be of sufficient size and type to meet the bearing capacity of the soil. This unit covers the various types of footings and foundations frequently used in building construction, with particular attention to the details of reinforced concrete piers and to footings for supporting steel columns. The use of piling with concrete footings where the soil has a low load-bearing capacity is also examined.

TECHNICAL INFORMATION

Footings. The fundamental supporting elements designed to carry the total weight of the building. They must be constructed upon undisturbed (virgin) soil and they must be calculated to support the load upon whatever type of soil is found on the site. They must always be below the frost line, otherwise the alternate expansion and contraction of the earth during freezing and thawing may heave the footing, the foundation wall, and the structure above.

Foundation walls. That part of a building which is located between the footing at the bottom and the supports of the floor nearest the grade at the top. Foundation walls provide (a) protection against water and frost penetration, (b) lateral support for the enclosed building area against pressures exerted by the surrounding earth, and (c) the distribution of the vertical loads of the superstructure to the footings.

Piles. Vertical supporting members driven into the ground to help bear the vertical load of any structure resting upon it. The basic principle is that any length of material driven into any substance other than water develops friction along its length. It is this friction that provides support. The total friction developed by the piles in the soil should always be greater than the load imposed upon the footings at the top of the pile. Piling therefore is one method of increasing the load-carrying capacity of any type of soil to the degree required for a structure.

Types of Footings

Mat footing. This serves as a foundation for the entire building;

REINFORCED CONCRETE SLAB ACTS AS ONE UNIT AND DISTRIBUTES BUILDING LOADS OVER THE ENTIRE SURFACE

Figure 6-1. Mat Footing

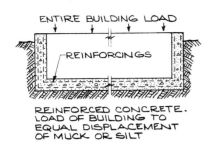

Figure 6-2. Raft or Boat Footing

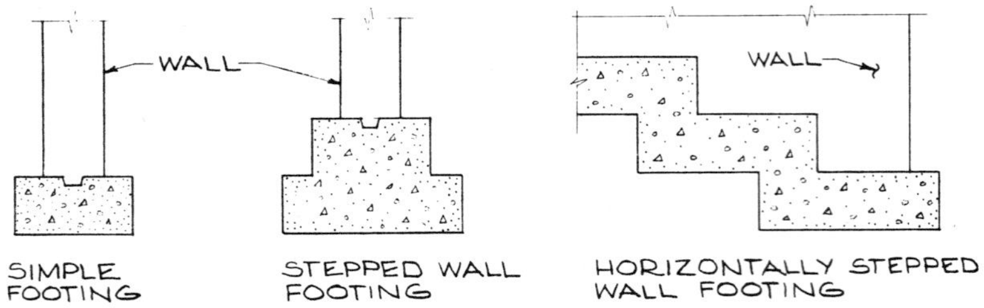

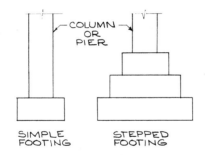

Figure 6-3. Wall Footings

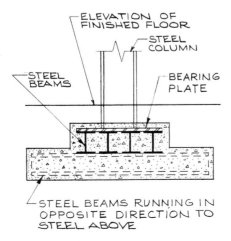

Figure 6-4. Column or Pier Footings

ELEVATION OF FINISHED FLOOR

STEEL COLUMN

STEEL BEAMS

BEARING PLATE

STEEL BEAMS RUNNING IN OPPOSITE DIRECTION TO STEEL ABOVE

Figure 6-5. Grillage Footing

i.e., it is both footing and foundation wall (see Figure 6-1). Made of reinforced concrete, it acts as one unit, distributing the building loads over its entire area. Because of its proximity to the ground surface, it is generally used in areas where there is no freezing.

Raft or boat footing. This type (see Figure 6-2) is used in areas where quicksand, muck, or silt are the bearing soils. It is made of reinforced concrete and, most important, its construction must be watertight. Its size is dictated by the total building load in relation to displacement of the muck and silt upon which it rests.

Wall footing. Usually made of concrete or reinforced concrete, it is designed to spread the imposed loads of a foundation wall over a larger ground area. The spread may be designed as a series of steps in a vertical or horizontal direction (see Figure 6-3).

Column pier footing. This type is used under individual columns or piers. The columns may be of brick, concrete block, reinforced concrete, or steel. As in wall footings, the column footing is designed to spread the load of the column over a larger ground area. This can be done in a number of ways, as shown in Figure 6-4 for a simple and a stepped footing, and in Figure 6-5 for a grillage footing. The grillage is constructed of a series of steel beams placed adjacent and parallel to each other, with a second layer of similar steel placed at right angles. A bearing plate, which

will receive the steel column, is anchored to the grillage. The entire grillage area is then encased in concrete. This type of footing is used for columns supporting very heavy loads.

Steel column on reinforced footing. The concrete footing (see Figure 6-6) is reinforced with 14 @ #5 rods each way, or a total of 28 reinforcing rods. (*Note:* The #5 represents the rod diameter in eighths of an inch.) In the footing are two 3/4-inch anchors installed to receive the setting plate. The 1/4-inch setting plate is placed true and level on cement grout before the billet plate is set. Then the steel column with a billet plate welded to its base is placed on the setting plate and bolted down. Locations and sizes of reinforcings are given only for illustrative purposes. Steel column, billet plate, and footing sizes are calculated to meet the imposed loads and soil-bearing capacity.

Steel column on reinforced concrete pier or column. The reinforced concrete column (see Figure 6-7) is shown resting on a reinforced concrete footing. Four vertical reinforcing rods in the concrete pier or column are bound with lateral ties, usually spaced 11 to 12 in. apart to keep the verticals rigid while the concrete is poured into the forms. To create a stronger bond between footing and pier or column, there are metal rods or dowels protruding from the footing that tie it to the column when it is poured. On top of the pier or column is the grout, the setting plate, and the steel column with the billet plate welded to the bottom. Reinforced concrete pier and column schedules on the steel plans give the minus dimensions; that is, the location of the top and bottom in relation to elevation 0'-0", which in most cases represents the level of the finished first floor.

Concrete pier and column reinforcing. Reinforced concrete piers or columns may have 4, 6, 8, 10, and more vertical bars, depending on the size of the column. Figure 6-8 indicates such columns with their vertical bars and lateral ties as required by the American Concrete Institute. The 4-bar column requires a single tie per set. A set occurs every 11 or 12 in. of column height.

Foundations

Foundation walls. The most commonly used types of foundation walls are shown in Figure 6-9. Poured stone concrete walls and concrete block walls are used for relatively light construction such as houses and smaller commercial or industrial buildings. The reinforced concrete wall, with both vertical and horizontal reinforcing rods to resist both vertical and lateral stresses, is designed for larger and heavier structures.

Corner piers. Figure 6-10 illustrates two design variations for corner pier foundations that are to receive the structural steel columns. At (A) the reinforced concrete pier has its individual footing designed to carry the weights imposed. Such footings are often larger and deeper than the adjacent wall footings. At (B) the corner pier is an integral part of the foundation wall, and the footings for both wall and pier are of the same thickness. It can safely be assumed that the first type (A) can carry heavier loads than type (B).

Types of Piles

Wood piles. Tree trunks of Douglas fir or Southern pine are driven with the small end down (see Fig. 6-11, A). The diameter at the tip varies from 6 to 8 in., while the butt end is rarely less than 12 in. The pile is tapered uniformly and the wood is untreated if embedded below permanent ground-water level. Such untreated timber piles

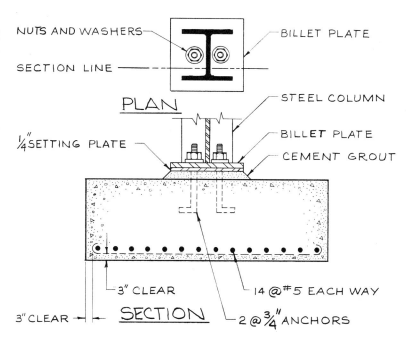

Figure 6-6. Steel Columns on Reinforced Concrete Footing

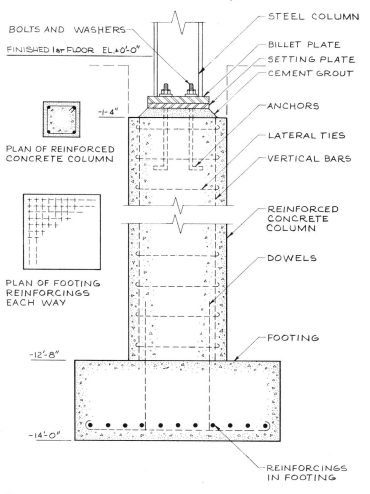

Figure 6-7. Steel Column on Reinforced Concrete Column on Reinforced Concrete Footing

may be considered permanent if entirely embedded in soil and cut off below the lowest present or future ground-water level or if entirely submerged in fresh water. Where unfavorable conditions occur such as ground water containing alkali or acids, untreated piles cannot be used. Piles for these conditions are impregnated with coal tar or other preservative to stop them from rotting.

Precast concrete piles. These may be round, square, or octagonal, and are made up to 40 ft. in length (see Figure 6-11, B). The piles are reinforced internally to resist stresses caused by handling and driving.

Cast-in-place piles. These are constructed by driving a steel shell into the ground by means of a heavy core called a mandrel (see Figure 6-11, C). When sufficient resistance has been obtained, the core is withdrawn. The driven shell is then inspected and filled with concrete. Such piles are about 37 ft. in length.

Metal pipe piles. The piles are cylindrical shells 10 to 18 in. in diameter. The thickness of the steel shell varies from 1/4 to 1/2 in. (see Figure 6-12). Sections are sunk in 20-ft. lengths. This type of pile is usually driven to a point of refusal or to rock bottom. As it is driven down, the pipe fills with soil which is then blown out by means of compressed air. When point of refusal or rock is reached, the pipe is filled with concrete. Pipe lengths of up to 100 ft. have been used.

Structural steel H-column piles. The steel H-columns are usually driven to a point of refusal or rock bottom (see Figure 6-13). They can be driven through dense, hard earth and can support very heavy loads. The top of the steel H-column is encased in concrete and coated with a rust-inhibiting paint to about 8 in. above and 2 ft. below grade.

Pile clusters. Piles are generally driven in clusters (see Figure 6-14) specifically designed to support columns, piers, and walls. A cluster may include from 3 to 25 piles and more if necessary to meet load conditions.

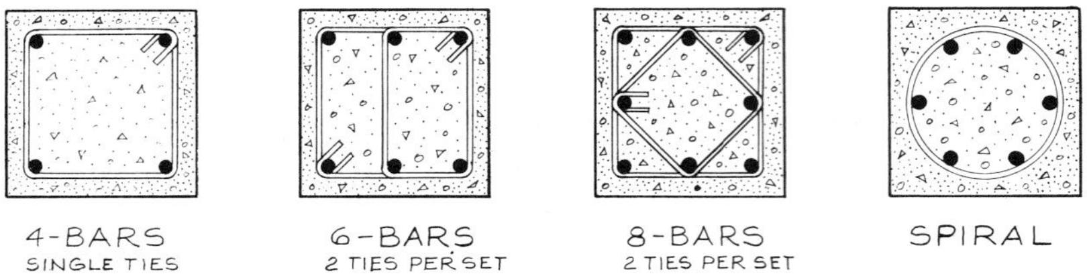

![4-BARS SINGLE TIES | 6-BARS 2 TIES PER SET | 8-BARS 2 TIES PER SET | SPIRAL]

Figure 6-8. Typical Arrangement of Reinforcing Steel in Reinforced Concrete Columns

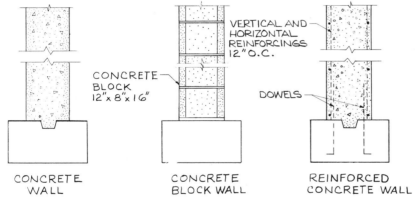

Figure 6-9. Typical Foundation Walls

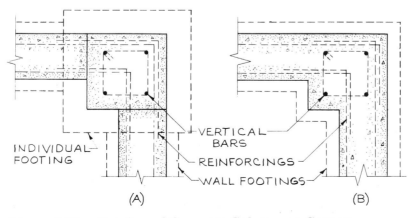

Figure 6-10. Reinforced Concrete Columns at Corners

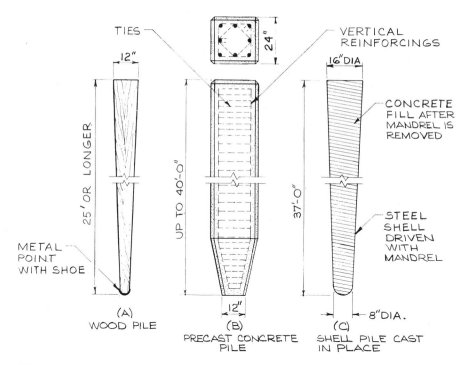

Figure 6-11. *Piles*

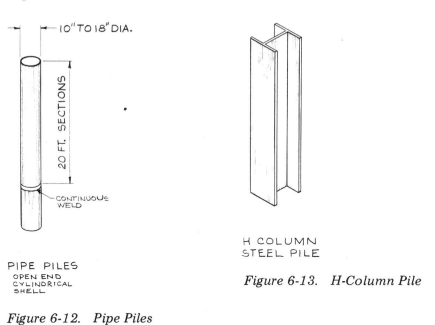

Figure 6-12. *Pipe Piles*

Figure 6-13. *H-Column Pile*

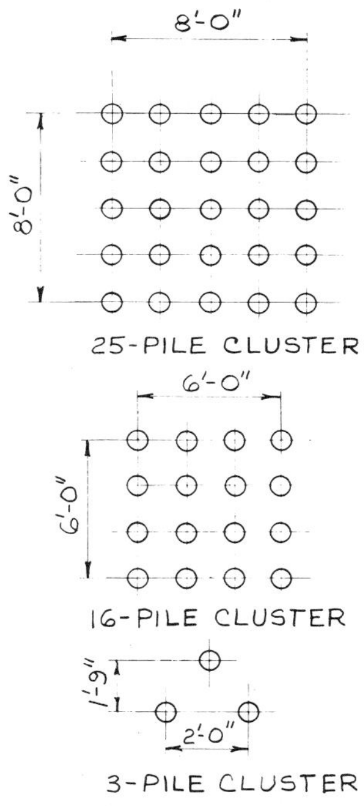

Figure 6-14. *Pile Cluster*

1. What word means a structural unit used to distribute building loads to the bearing materials?

2. What are the different types of footings?

3. What are the dimensional variations of the diameter of the tips of wood piles?

4. Name two kinds of wood used for wood piles.

5. Name two types of concrete piles.

6. The steel casing for a shell pile is driven into the ground with the aid of what?

7. The diameter of metal pipe piles have what dimensional variation?

8. Sections of the metal pipe pile are sunk in lengths of how many feet?

9. After the metal pipe piles are in place and the soil inside the pipe is blown out by means of compressed air, the pile is filled with what?

10. What is the range in the number of piles in a pile cluster?

11. The setting plate receives what type of plate?

12. The setting plate is set into a bed of what?

13. What is the diameter of a #5 reinforcing bar?

14. The steel column stands plumb and true on what type of plate?

15. The base of a steel column and the billet plate are what?

16. Foundation walls may be constructed of what materials?

ASSIGNMENT

1. Prepare a sketch of a reinforced concrete column resting on a reinforced concrete footing. Place a steel column on top indicating the grout, setting plate, and billet plate.

Show all reinforcing.

2. Sketch three types of piles.

3. Sketch three types of wall footings.

SUPPLEMENTARY INFORMATION

Reinforcing

See Table 6-1 for standard sizes and weights and new bar designation numbers. Reinforcing rods or bars are designated by a symbol (#)

and a number, for example: 4 - #2 @ 12″ OC, which means four 1/4-in. round rods at 12 in. on center.

Spacing of reinforcings. The space between reinforcing rods must satisfy the conditions shown in Figure 6-15. When concrete is poured over the reinforcing rods within a footing, the larger particles of the aggregates must flow around and beneath the reinforcing rods so that no voids are created. Rods must be entirely surrounded by the concrete if their full efficiency is to be realized.

Edge distance of reinforcing rods is affected by aggregate size, size of

Table 6-1. *Standard Sizes and Weights of Reinforcing Bars*

Former Bar Designation (in.)	New Bar Designation	Unit Weight (lb. ft.)	Diameter	Cross Section (sq. in.)
1/4 round	#2	0.167	0.250	0.05
3/8 round	#3	0.376	0.375	0.11
1/2 round	#4	0.668	0.500	0.20
5/8 round	#5	1.043	0.625	0.31
3/4 round	#6	1.502	0.750	0.44
7/8 round	#7	2.044	0.875	0.60
1 round	#8	2.67	1.000	0.79
1-1/8 round	#9	3.4	1.128	1.00
1-1/4 round	#10	4.303	1.127	1.27
1-3/8 round	#11	5.313	1.410	1.56

bar or rod, and by moist or wet ground conditions (see Figure 6-16).

Purpose of reinforcing bars. What is the exact function of the reinforcing in footings and foundations as shown in several illustrations in this unit? Reinforcing bars are placed near the bottom of footings and slabs and in walls near the edges of the wall. To understand the reason for the location of the bars, assume a board supported at both ends on two carpenter's horses. If a man were to stand on the board it would bend under his weight (see Figure 6-17). The bottom of the board would be in tension (pulling apart) while the top would be in compression. Similarly, a concrete footing supporting a foundation wall acts as the loaded board. The reinforcing bars, strong in tension, will not be stretched, hence preventing (counteracting) the bending action of the footing. The same principle applies to a wall with vertical bars near its outside and inside surfaces—here the tensile strength of the bars prevents the loaded wall from buckling because of the downward pressures of the imposed weight of the building. A wall that would tend to buckle toward the inside because of the pressure of the earth on one side would have the vertical bars near the inside surface.

Loads imposed on concrete floor slabs require the reinforcings near the bottom of the slab. Slabs subjected to hydrostatic pressure (lifting of the slab) would require bars near the top of the slab.

Testing the Load-Bearing Capacity of a Concrete Pile

Where there is doubt as to the true carrying capacity of a driven pile, load tests must be performed on such piles. The method of testing and the equipment used is illustrated in Figure 6-18, A, B, C. The steps in erecting the equipment is outlined in the following:

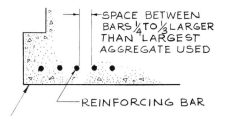

Figure 6-15. *Bar Spacing*

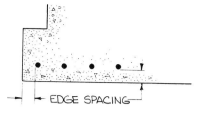

Figure 6-16. *Bar Edge Distance*

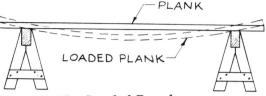

Figure 6-17. *Loaded Board*

Table 6-2. *Schedule of Loading*

Increment Numbers	Loading		Notes
	% Design Load	Actual Load	
0	0	0 tons	Record initial reading of all dials, scales, and bench marks
1	50	65.0 tons	Apply first increment of load and record settlements ⊕
2	75	97.5 tons	Apply second increment of load and record settlement ⊕
3	100	130.0 tons	Apply third increment of load and record settlement ⊕ ▲
4	125	162.5 tons	Apply fourth increment of load and record settlement ⊕ ▲
5	150	195.0 tons	Apply fifth increment of load and record settlement ⊕ ▲
6	175	227.5 tons	Apply sixth increment of load and record settlement ⊕ ▲
7	200	260.0 tons	Apply seventh increment of load and record settlement ⊕ ▲ ●
5	150	195.0 tons	Reduce load to fifth increment and record readings—hold for 1 hr.
3	100	130.0 tons	Reduce load to third increment and record readings—hold for 1 hr.
1	50	65.0 tons	Reduce load to first increment and record readings—hold for 1 hr.
0	0	0 tons	Reduce load to zero and record readings; leave all dials and gages intact for 24 hrs.
0	0	0 tons	Record final readings and dismantle load test box and equipment

⊕ Under each load increment, settlement observations shall be made and recorded at 1/2 min., 1 min., 2 min., 4 min., and each 4 min. thereafter after application of load increment, except in the instance of the total load where, after the 4-min. reading, the time interval shall be successively doubled until the final settlement limitation is reached and the load removed.

▲ After the proposed working load has been applied and for each increment thereafter, the test load shall remain in place until there is no measureable settlement (0.002 in. or less) in a 2-hr. period.

● The total test load shall remain in place at least 96 hrs., and until settlement does not exceed 0.00 in. in the last 48 hrs.

Procedure

1. Level and prepare subgrade in the vicinity of the test pile.

2. Position telltale extensometer, bearing plates, and hydraulic jacks concentrically over the pile head.

3. Carefully construct the load test box (see Figure 6-18, A) to provide a concentric load on the pile.

4. Load test box with sufficient soil to provide required reaction for testing.

5. Establish two reference bench marks for level readings.

6. Position reference beam on its own independent supports.

7. Attach all gages and scales as shown in Figure 6-18, B and C.

8. Give ample notification, prior to the start of the testing, to the designated representative of the Building Department.

9. Using the hydraulic pump, induce load onto pile through jack acting against the test box, making certain to maintain constant load at each increment (Table 6-2).

10. Record data and proceed as outlined in the Schedule of Loading.

11. A complete record of the load test shall be filed with the Building Department (the commissioner or other designated head).

The test pile must be instrumented so that the movements of the base and butt can be independently determined. The extensometers must provide readings to the nearest 0.001 in. Additional settlement observations shall be taken using an engineer's level reading to 0.001 ft.

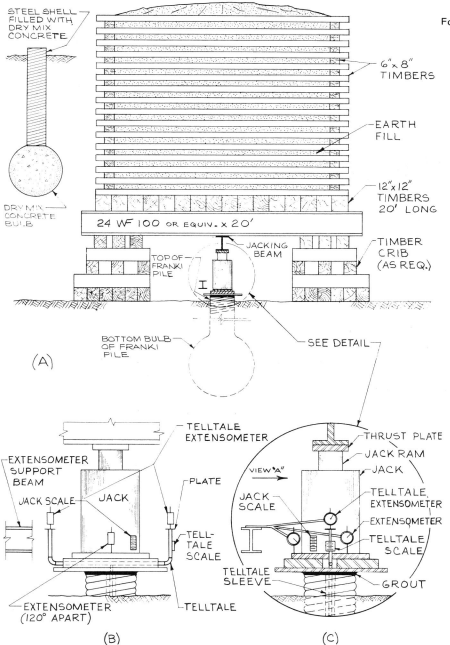

STEEL SHELL FILLED WITH DRY MIX CONCRETE

DRY MIX CONCRETE BULB

6"x8" TIMBERS

EARTH FILL

12"x12" TIMBERS 20' LONG

TIMBER CRIB (AS REQ.)

24 WF 100 OR EQUIV. x 20'

JACKING BEAM

TOP OF FRANKI PILE

SEE DETAIL

BOTTOM BULB OF FRANKI PILE

(A)

TELLTALE EXTENSOMETER

EXTENSOMETER SUPPORT BEAM

JACK SCALE

JACK

PLATE

TELL-TALE SCALE

EXTENSOMETER (120° APART)

TELLTALE

(B)

VIEW "A"

THRUST PLATE

JACK RAM

JACK

JACK SCALE

TELLTALE EXTENSOMETER

EXTENSOMETER

TELLTALE SCALE

TELLTALE SLEEVE

GROUT

(C)

Courtesy of the Frankie Pile Corp.

Figure 6-18.   Load Test Procedure

7

Site and Building Drainage

INTRODUCTION

This unit deals with the drainage of water from roof areas and the ground areas surrounding the building, which includes yard and area drains, parking area drains, and subsoil drains.

TECHNICAL INFORMATION

Rainwater from roofs, ground surface water, and water from other areas must be collected and conducted to some point of discharge. In urban areas it is discharged into a storm sewer or a combination storm and sanitary sewer by means of the house drain and house sewer. In suburban areas it is discharged into storm sewers or combination storm and sanitary sewers if they are available, and if not, into underground basins which receive the rainwater and in turn slowly dissipate it into the earth. In rural areas it is either discharged directly to the ground or piped to underground basins, lakes, streams, and rivers.

Roof Drainage

There are two general methods of draining a roof: by interior ridges and valleys built into the roof, and by eave gutters under overhanging eaves. Eave gutters are used for gable roofs or other types having a steep slope and overhanging eaves such as are found on private dwellings, barns, and similar structures (see Figure 7-1). Interior ridges and valleys are used for structures with flat roofs (see Figure 7-2).

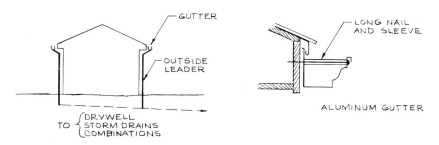

Figure 7-1. *Gutters for Drainage of Pitched Roofs*

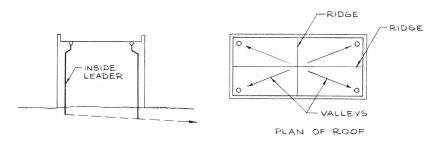

Figure 7-2. *Roof Gutters or Valleys*

49

Leaders. The water is conducted from the roof to the point of discharge by pipes called leaders, which may run inside or outside the

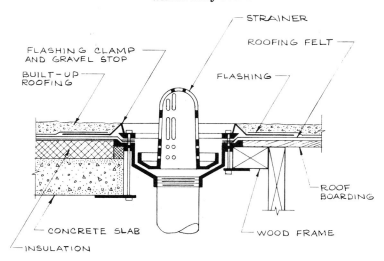

Figure 7-3. Roof Drain Detail

Table 7-1. Roof Area Drained by Various Sizes and Types of Outside Leaders

Type of Leader	Size of Leader (in.)	Cross Sectional Area of Leader (sq. in.)	Roof Area Drained (sq. ft.)
Plain round	3 diameter	7.07	1060
	4 diameter	12.57	1885
	5 diameter	19.63	2945
	6 diameter	28.27	4240
Corrugated rectangular	1 3/4 × 2 1/4	3.9	585
	2 3/8 × 3 1/4	7.71	1156
	2 1/4 × 4 7/8	10.96	1644
	3 3/4 × 5	18.75	2812
	2 × 3	6	900
	2 × 4	8	1200
Plain rectangular	3 × 4	12	1800
	4 × 5	20	3000
	4 × 6	24	3600

Table 7-2. Size of Gutters and Leaders (rate of 4" rainfall per hour)

Area of Roof to be Drained (sq. ft.)	Size of Gutter (in.)	Diameter of Inside Leader * (in.)
up to 90	3	1 1/4
91 to 270	4	2
271 to 810	4	2 1/2
811 to 1800	5	3
1801 to 3600	6	4
3601 to 5500	8	5
5501 to 9800	10	6

*Outside leaders should be one pipe size larger.

building. Inside leaders are of cast iron or wrought iron, bell-and-spigot type, and must be made gas- and watertight by using hemp (oakum) and lead. Where leaders pass through the roof, the opening must be tightly flashed to prevent leakage around the opening (see Figure 7-3). Outside leaders may be any one of several types of sheet metal such as galvanized iron, copper, stainless steel, or aluminum. They are available in various standard and custom-designed shapes such as round, square, or rectangular.

Size of gutters and leaders. The size of gutters and leaders required is governed by the area of roof to be drained and the intensity of rainfall. This of course varies in different parts of the country. Rainwater from roofs must be carried away as fast as the rain falls; therefore the maximum rate of rainfall that occurs during heavy thunder showers must be the governing factor. In some instances a rate of fall of as much as 8 in. per hour has been recorded. Gutters and leaders large enough to carry away this amount of water must be installed.

The first step in designing a system of gutters and leaders is to locate the leaders. This may be dictated largely by architectural conditions. However, the maximum spacing recommended for leaders is 75 ft. The next step is to calculate the area drained per leader, and from this to determine the size of the leader. Authorities differ in their recommendations, which range from 75 to 250 sq. ft. of roof surface to each sq. in. of leader cross section. This difference is reflected in Tables 7-1 and 7-2. The Copper and Brass Research Association recommends that 150 sq. ft. of roof area be allowed for each sq. in. of leader area. Table 7-1 shows the relationship of size and shape (type) of leader to roof area drainage.

For semicircular sheet metal gutters under overhanging roofs, the top dimension (diameter) of the gutter is equal to the size of the leader. Other shapes should have the same cross-sectional area.

The sizes of gutters and leaders must also be correlated. Table 7-2 gives the size of each for various roof areas, as recommended in a United States Department of Commerce Report based on a rainfall rate of 4 in. per hour.

Yard and Area Drains

The purpose of yard and area drains (see Figure 7-4) is to convey the storm water as fast as it falls, thus preventing the formation of pools of water.

In paved yards such as courts, schoolyards, and other similar places, *drainage inlets* or *catch basins*, the more generally used term, are built below the surface (see Figure 7-5). The catch basin may be of brick, concrete block, concrete, or precast concrete. A screened opening in the top permits the rainwater to enter the catch basin from which it flows by drainpipe to the storm sewer, combination storm and sanitary sewer, or underground basins which disperse the water back into the ground. The opening must be screened in order to keep out material carried by the water such as paper, leaves, or sticks that might clog the drainpipe. All paved areas, yards, courts, etc. exceeding 15 sq. ft. should be drained into the storm or combination storm and sanitary sewer, if such is available, or into the underground basins.

The capacity of yard and area drains depends on the rate of rainfall. This may be estimated at 4 in. per hr. Thus an area of 60 ft. × 100 ft. to be drained could yield the following quantity of water:

60 × 100 = 6000 sq. ft.
6000 × 4/12 = 2000 cu. ft. per hr.
= 33.3 cu. ft. per min.
= 249.45 or 250 gals. per min.

Note:

1 cu. ft. = 7.491 gals.

Drainage within the Building

Where cellar floor drains are located below the level of the public storm or combination storm and sanitary sewers in the street, sump pumps are employed to lift the water to higher levels so that it can

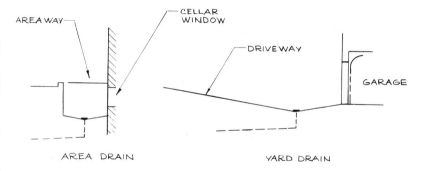

Figure 7-4. *Area and Yard Drains*

be drained off (see Figure 7-6). A detail of the sump pump is shown in Figure 7-7.

Floor drains in cellars must include a trap of the deep-seal type (see Figure 7-8). Such drains should be provided with a water supply by means of a faucet located not more than 3 ft. above the floor, except where other suitable means for maintaining the trap seal are provided.

Subsoil Drainage

Underground basins. This type of basin is manufactured in three sections: drain ring, solid ring, and dome (see Figure 7-9). These underground basins are used singly or in series to collect rainwater on paved areas. The water runs either into the opening at the top of the dome or is collected in a series of yard drainage inlets (catch basins) which are connected by pipes to the drainage basin (see Figure 7-10). The lower ring of the underground basin is provided with drain hole openings designed especially to let the water dissipate into the earth without allowing the holes to become clogged with earth. Drainage also takes place at the bottom of the basin.

Sizes. Drain rings and solid rings are obtainable in 4-, 6-, 8-, and 10-ft. diameters. The available heights of drain rings and solid rings are 2'-0" and 4'-6".

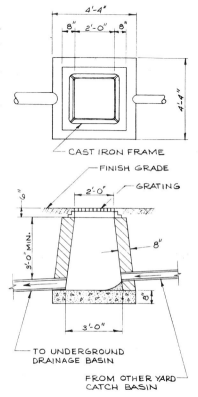

Figure 7-5. *Yard Drainage Inlet or Catch Basin*

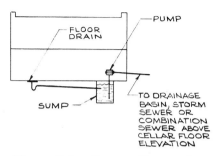

Figure 7-6. *Sump Pump Installed in Cellar Floor*

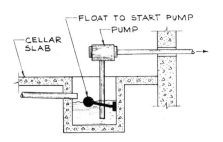

Figure 7-7. *Detail of Sump Pump Servicing Floor Drain or Drains*

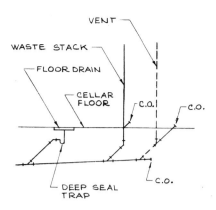

Figure 7-8. *Floor Drain*

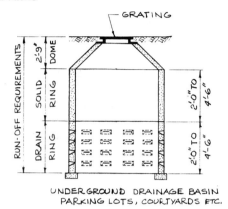

UNDERGROUND DRAINAGE BASIN
PARKING LOTS, COURTYARDS ETC.

Figure 7-9. *Underground Basin*

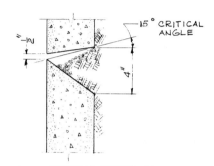

OPENING DETAIL IN DRAIN RING

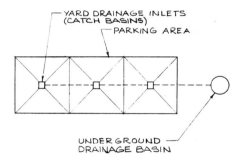

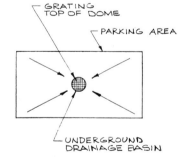

Figure 7-10. *Parking Area Drainage*

Foundation Drains

The foundations and cellar floors of buildings are frequently provided with exterior drainage systems (foundation drainage) designed to carry off ground water and surface water which penetrate through backfill (see Figure 7-11). The simplest form of subsoil drainage for this condition consists of a line of open-joint clay tile laid along the outside perimeter of the building in gravel backfill at the level of the bottom of the footings (see Figure 7-12). The water collected by the pipes is carried into a dry well located at least $10'-0''$ away from the building. This system is primarily designed to drain off surface water which penetrates through backfill and would otherwise build up against the foundation walls and enter into the cellar. A dry well built of field stone, gravel, and sand is shown in Figure 7-13. This type is used in typical rural areas. Wherever one builds there are city, town, and state requirements governing building and site drainage, and plans must be submitted for official approval. The following requirements and sample problem are based upon the code requirements of a typical town in the United States.

Drainage Requirements

The following standards must be used for establishing site drainage of projects submitted for approval to the planning board and town engineer:

1. All storm water shall be contained on the site.

2. Where soil conditions and all other requirements of the town engineer so justify, an underground basin shall be selected in accordance with the following formula:

A. *Water runoff factors:*

1. Pavement (graded 5% or less)   0.90
2. Pavement (graded in excess of 5%) 0.95
3. Roof deck   1.00
4. Seeded (graded 5% or less)   0.08
5. Seeded (graded in excess of 5%) 0.17
6. Other areas (minimum)   0.30

3. Computations for cu. ft. of rainfall shall be determined as follows:

A. Required capacity (cu. ft.) of area (sq. ft.) to be drained *times* runoff factor *times* 0.12. (1-1/2 in. of rainfall = 0.12 ft.).

4. Underground basins with drainage

slots shall be designed allowing for ground absorption as follows:

A. Total capacity (cu. ft.) of area in sq. ft. *times* runoff factor *times* 0.12.

5. Determination for selection of underground basins may be guided as follows:

A. Standard 8'-0" diameter of basin contains 42 cu. ft. for each ft. of height. Standard 10'-0" diameter contains 67 cu. ft. for each ft. of height.

B. No credit for drainage shall be given for basin height above the level of inlet pipes.

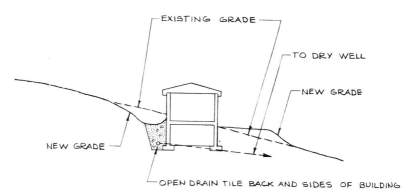

Figure 7-11. Foundation Drainage

Sample problem

A. *Drainage Schedule*

B. *Capacity* equals 12,740 sq. ft. (required areas *times* runoff factors) *times* 0.12 = 1,528.8 cu. ft., total capacity required.

C. Total lin. ft. of underground basins equals

$$\frac{1528.8}{42 \text{ cu. ft./ft.}} = 36.40 \text{ lin. ft.}$$

D. Select 4'-6" high reinforced sections

$$\frac{36.4}{4.5} = 8.1, \text{ or 8 sections (4'-6" high and } 8'-0" \text{ dia.)}$$

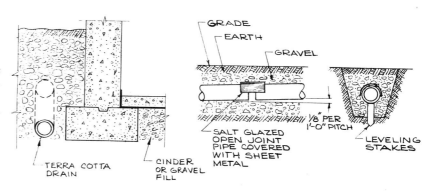

Figure 7-12. Details of Foundation Drain

E. Total linear feet of underground basins (assume 10 ft. dia. rings) equals:

$$\frac{1528.8}{67} \text{ cu. ft./ft.} = 22.82 \text{ lin. ft.}$$

F. Select 4'-0" high reinforced sections

$$\frac{22.82}{4.0} = 5.7, \text{ or 6 sections (4'-0" high and } 10'-0" \text{ dia. rings).}$$

G. The grades of the property will determine location of underground basins.

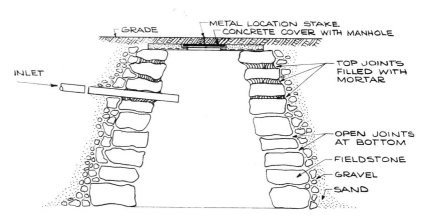

Figure 7-13. Fieldstone Dry Well

Area	Use	Material	Runoff Factor	Drainage
6,500 sq. ft.	parking	paving	0.90	5,850 sq. ft.
6,500 sq. ft.	building	roof deck	1.00	6,500 sq. ft
3,000 sq. ft.	landscaping	seeded	0.08	240 sq. ft
500 sq. ft.	undeveloped	earth	0.30	150 sq. ft.
16,500 sq. ft. site total		equivalent drainage		12,740 sq. ft.

1. In urban areas, where may rainwater be discharged directly?

2. Eave gutters are used on what type of roofs?

3. Inside leaders must be made of what material?

4. What is the size of gutters and leaders governed by?

5. What is the recommended spacing of leaders?

6. The Copper and Brass Research Association recommends how many sq. ft. of roof area to be allowed for each 1 sq. in. of leader area?

7. What must floor drains in cellars have?

8. What is the water runoff factor for a roof deck?

9. For each 1 ft. of height, an 8-ft. diameter underground basin contains how many cu. ft.?

10. For each 1 ft. of height, a 10-ft. diameter underground basin contains how many cu. ft.?

ASSIGNMENT

1. Determine the number of underground basins of 8-ft. diameter required for the following paved areas:

12,500 sq. ft. for parking
7,200 sq. ft. for roof deck
2,000 sq. ft. for landscaping

SUPPLEMENTARY INFORMATION

Ground-Water Level

When the ground-water level is found to be considerably above the lowest floor of the building, it is necessary to waterproof the floor and walls and to design the structure to resist the pressure of the water (see Figure 7-14).

For each foot of height the pressure is 62-1/2 PSF (pounds per sq. ft.). If the water rises 3 ft. above a basement floor, the upward pressure on the floor is 3 times 62-1/2, or 187-1/2 PSF. This is far more than the weight of the usual 4-in. thick concrete slab, which weighs approximately 45 PSF — therefore the slab must be heavier and reinforced against this pressure. Correspondingly, lateral pressure is exerted wherever ground water collects against foundation walls, the pressure varying with the height of the water above the level under consideration. The foundation walls therefore must be reinforced in order to withstand this lateral pressure.

A proper method of foundation waterproofing is shown in Figure 7-15, Detail A. The waterproof membrane should be protected by

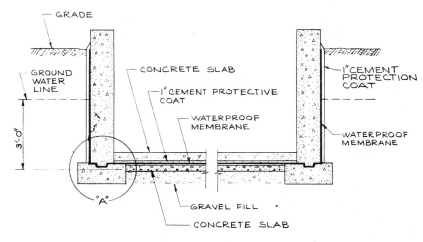

Figure 7-14. Membrane Waterproofing

1-in. cement or 1/2-in. wallboard or by 4-in. brick. The floor membrane should be placed on a leveling bed of concrete having a lightweight aggregate and of 1-in. minimum thickness.

Maintaining Ground-Water Level

To prevent the ground-water level from rising higher than desired, a sealed tank with a pipe gooseneck is employed as shown in Figure 7-16. The terra cotta drain tile adjacent to the footing of the building collects the ground water, which is conducted to a sealed concrete pit with a gooseneck pipe emptying into a sump pit where the water is pumped out. The location of the top of the gooseneck determines the level at which the ground water must rise no further.

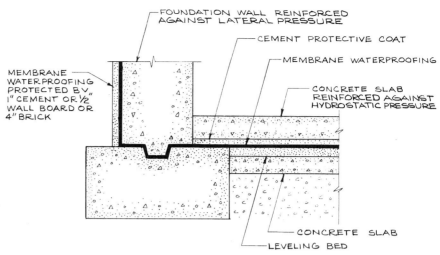

Figure 7-15. Detail "A"

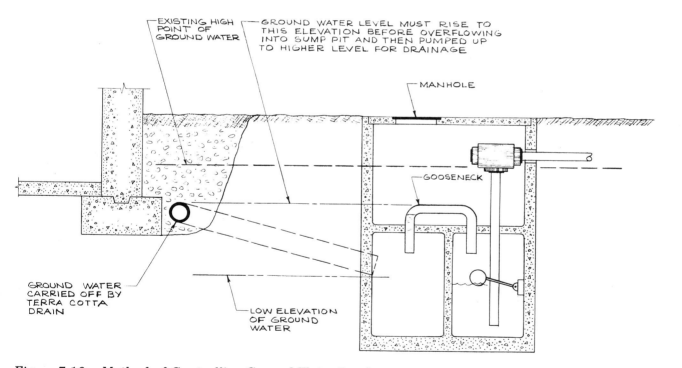

Figure 7-16. Method of Controlling Ground Water Level

8

Cementitious Materials

INTRODUCTION

This unit deals with the study of cementitious materials such as lime, gypsum, cement, plastic, and asphalt and their combination with other materials to form plasters and concrete mortars used for finishing, paving, and bonding of masonry blocks and other building units.

TECHNICAL INFORMATION

Lime

Lime is the product resulting from the crushing of dolomite limestone, a rock consisting mainly of magnesium and calcium carbonate, and the burning of it in kilns to 2000° F., causing the limestone to break down into carbon dioxide (gas) and calcium oxide (solid). The solids are pulverized into a fine powder known as quicklime.

Quicklime can never be used as such for structural purposes; it must first be slaked in water to form a putty. During the slaking process, violent, almost explosive reactions occur in which very large quantities of heat are given off. The quicklime is marketed in paper bags, steel drums, and in bulk (see Figure 8-1).

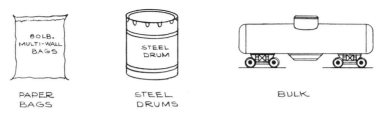

Figure 8-1. Containers for Marketing Quicklime

Hydrated lime is lime that has gone through the slaking process; that is, enough water has been added to satisfy its chemical affinity for water, after which it is air-separated and the solids sifted to a fine dry powder so that 98% will pass through a 200-mesh screen. Today very little slaking of lime is done at construction sites; instead, manufacturers produce hydrated lime, which is marketed in various forms as shown in Figure 8-2. Hydrated lime is used for plastering, as well as in cement mortars to increase their workability, plasticity, and water-holding capacity. The lime, further, reduces the separation or segregation of the sand and cement in mortars or plasters.

Gypsum

Gypsum is a hydrated sulfate of calcium occurring naturally in sedimentary rocks, which is used in

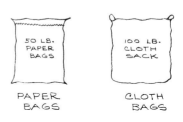

Figure 8-2. Containers for Marketing Hydrated Lime

making plaster of Paris. When heated from 325 to 340° F., it loses most of its water and is then finely ground into powder called plaster of Paris. The heating process is known as "calcining." This calcined product is used for plasters, plaster-type products and materials, fireproofers, and for slowing down the setting time of portland cement.

Cement

The materials necessary in the manufacture of cement are lime, silica, iron oxide, and alumina, which are obtained by mixing an impure clay-bearing limestone with pure limestone, or by mixing limestone with clay or shale. These ingredients are mixed in proper and closely controlled proportions. During the manufacturing process, the raw materials are ground and burned to form "clinkers." These are then cooled and pulverized to produce cement. The strength of the cement is increased with the fineness of grinding.

Portland cement finds its greatest use in building construction for making concrete and mortars. Concrete may be reinforced, cast, or prefabricated into numerous building units, while the mortars serve as the bonding agent in masonry work. When water is added to portland cement, a paste is formed which soon loses its plasticity and begins to harden. The initial set takes place within 45 minutes and a final set in about 10 hours. Portland cement continues to harden increasingly over a very long period of time.

There are five important types of portland cement generally used in building construction:

1. Standard portland — for general use.
2. Modified portland — slow setting and less heat generated.
3. High early strength portland — where quick setting and early strength are necessary.
4. Low heat portland — very slow setting and very little heat generated.

5. Sulfate-resisting portland — where ground water and soils will attack other types of portland cements.

Other important uses of portland cement are in the manufacture of asbestos-cement products, cement brick, terrazzo and artificial stone; as a basis for paints for concrete and masonry materials; and as a plaster for exterior surfaces. Some minor uses are for grouting, pointing, floor finishes, fill for leveling, and numerous others. Portland cement is sold in bags weighing 94 lbs., and occupies 1 cu. ft. by volume. It is also sold in bulk.

Masonry cements are mixtures of portland cement with hydrated lime, silica, slag, and other additives. Because most masonry mortars are covered by patents, the exact make of the mixtures is not disclosed. There are two types of masonry mortars: Types I and II, which are available in white or gray color and in both waterproof and nonwaterproof.

Keene's cement is derived from gypsum. When gypsum is heated at a high temperature, between 1000 to 1400° F., all the water is driven off and the product, called hard-burned gypsum, is the result. By grinding and the addition of small quantities of alum, a tight, dense, water-resistant plaster is obtained.

Plastics—Synthetics

In the modern sense, plastics and synthetics can also be considered cementitious materials. They are used as adhesives, paints, and resilient flooring. When mixed with marble, cork, plastic, and other types of aggregate materials, they are used to produce seamless flooring and precast terrazzo-type flooring. Synthetics are also used with cement to make a masonry-type mortar.

Asphalt

Asphalt is a cementitious material, dark in color, consisting of bitumens, either natural or the residue left after the distillation of crude oil. Asphalt liquifies with heat and in petroleum distillates, and it can be emulsified with water. In building construction there are many asphalt products used, such as dampproofing, asphalt shingles and siding, asphalt flooring, caulking, expansion and pressure joint material, and built-up roofing.

The most important use of asphalt as a cementitious material is when it is combined with aggregates to make asphaltic cement paving for roads, driveways, parking lots, etc., and asphalt blocks for sidewalks, paths, and terraces.

Tar

Tar is obtained from coal as a by-product from making coke for the steel industry or from the distillation of wood. Although both tar and asphalt are similarly used in building construction, tar can withstand sunlight, water, cold, and heat better than asphalt, but it does soften considerably in direct radiant heat from sunlight in summer.

REVIEW EXAMINATION

1. What are the most commonly used cementitious materials?

2. Quicklime is the product resulting from the burning of what type of mineral?

3. What is lime that is slaked called?

4. In addition to its cementitious property, what other properties does lime contribute to mortar?

5. What is the material used to control the setting of portland cement?

6. Plaster and plaster-type products are made with what calcined mineral?

7. What is the cement that produces a hard, dense, water-resistant plaster?

8. Name the materials used in the manufacture of portland cement.

9. There are several types of portland cement. Name three.

10. The initial set of portland cement occurs in how many min. and the final set in about how many hrs.?

11. Portland cement is distributed in bags containing how many cu. ft. and weighing how many lbs.?

12. Masonry cements are used to make what two types of mortar?

13. Plastic cementitious materials are used to produce what two types of flooring?

14. The most important use of asphalt as a cementitious material is to produce what two things?

ASSIGNMENT

1. Make sketches of six construction items that employ cementitious materials.

SUPPLEMENTARY INFORMATION

Special Types of Portland Cement

Portland cement is also manufactured in several other types, of which two are used extensively in building construction: *white portland cement*, waterproof or nonwaterproof, used in mortars for stonework and to produce very white concrete; and *portland poz-*

zolanic cement, used in mortars to make more workable plastic mortars.

White portland cement is made by careful selection of suitable raw materials and by special manufacturing processes. Except for color, it has the same characteristics as gray portland cement and should be used in the same way. Because it is usually used where appearance is important, the forms should not be treated with grease, oil, or other materials that may cause stains.

Portland pozzolanic cement is made with portland cement and a pozzolana, a siliceous material. Various natural or artificial materials contain active silica, among them pozzola (volcanic ash), granulated slag, and pumice. Portland pozzolanic cement is produced by grinding together the portland cement clinker with the active silica until they are completely intermixed.

Oxychloric cement is a zinc or copper oxychloride and is used as the cementitious material to produce a very resilient, durable, seamless type of flooring.

Asphalt pavement. When asphalt-type pavement is to be installed, the type of soil (well-drained, slightly drained, or impervious to drainage) affects the sub-base course, which can vary from 0 to 8 in. thick depending on the type of drainage. Sub-base courses are of crushed stone or gravel, maximum size 2-1/2 in., well-rolled and tamped. Upon this sub-base is placed a base course, 4 to 5 in. thick of crushed stone or gravel of 1 in. maximum size, with fine aggregate applied on the surface to fill voids. This course should also be rolled and tamped. The final finish surface is of either hot or cold asphalt, 1 to 3 in. thick, depending on the type of wear by car or truck.

REVIEW EXAMINATION — UNITS 1 – 8

1. In order to determine the bearing capacity of the soil and subgrade conditions, what must be done?

2. Who makes the feasibility study for a large building or groups of buildings?

3. The Clerk-of-the-Works is generally employed by whom?

4. Does the Clerk-of-the-Works check all the materials to be used in the construction?

5. Undisturbed rock masses that are part of the original rock formation is the definition of what material?

6. What are detached rock particles 1/4 to 3 in. in size, rounded and waterworn?

7. What are noncoherent rock particles smaller than 1/4 in.?

8. The National Building Code requires that footings be placed at least how many feet below the frost line?

9. What is compressed and partially carbonized vegetable matter?

10. What is the bearing capacity in tons per square foot for hardpan?

11. What is the bearing capacity of soft clay?

12. The depth of footings is controlled by what?

13. What is the center load for the loading platform?

14. When the 50% overload is applied to the load on the platform, the settlement must not exceed what percentage of settlement under the proposed load?

15. A platform load of 6 tons in which dimension A = 5 ft. and dimension B = 8 ft., the center load is equal to how many lbs.?

16. What is the design load equal to in problem 15?

17. Why are sounding-rod tests made?

18. Rock drilling is necessary when rock is encountered in order to make certain that the boulder or thin layer of rock is not mistaken for what?

19. What are services such as sewage disposal and drainage systems, water, gas, electricity, and telephone, called?

20. Where are sewage disposal and drainage systems least likely to be found?

21. What is the portion of plumbing pipe leading from the building to the street sewer called?

22. What is the name of the vertical sewer pipe within the building that receives the discharge of a water closet?

23. A "fixture unit" has been set as a rate of discharge from a fixture at how many gals. per min.?

24. What is the rate of discharge of one water closet in terms of fixture units?

25. What are rural utilities usually limited to?

26. What is the purpose of the fresh air inlet immediately behind the house trap?

27. What is the purpose of vent lines in a plumbing drainage system?

28. In performing an absorption rate test, what must the measurements of the test pit be?

29. What is the length of a sub-soil sewage disposal drainage field serving 5 to 9 persons for a soil having a medium absorption rate?

30. When existing old buildings are to be removed for the construction of a new building, what is one of the first necessary steps?

31. What are the most frequently used types of temporary supports when the excavation of a new building is adjacent to an existing building?

32. In what position should shores be placed?

33. When bringing up the underpinning wall to within 2 or 3 in. of the bottom of the existing wall, what is the remaining space filled with?

34. What is the excavation required for the footing of a brick garden wall called?

35. When lagging boards and bracings are driven into the ground for deep excavations, what is the process called?

36. For what purpose are wellpoints driven into the ground?

37. Back-fill around walls after the forms are removed must consist of what material(s)?

38. The angle of slope, known as the angle of repose, for a retaining wall, is largely controlled by what factor?

39. What angle of repose is required for a soil composed of sand and clay?

40. The retaining wall is provided with weep holes how many ft. apart?

41. What is the type of footing most likely to be used where quicksand, muck, or silt are the bearing soils?

42. What is the pile that requires a steel shell known as?

43. What governs the size of gutters and leaders?

44. The capacity of yard and area drains depends on the rate of rainfall. How can this estimate be reached?

45. All paved areas should be drained into the storm or combination storm and sanitary sewer when such areas exceed how many sq. ft.?

46. The crushing of what material produces lime?

47. Where is the slaking process of hydrated lime accomplished?

48. How much time does it take for the initial set of portland cement?

49. Where quick setting and early strength is necessary, what type of portland cement should be used?

50. What material is tar obtained from?

9

Concrete

INTRODUCTION

Concrete is composed of sand, gravel, and other aggregates held together by a hardened paste of cement and water. A study in depth of this major construction material is beyond the scope of this book. This unit deals only with the design of concrete mixtures, the effect of the quantity of water on the strength of concrete, recommended water-cement ratios, the selection of consistency, and the selection of aggregates and other materials. It should be noted here that all concrete is now mixed by weight instead of by volume, as was done formerly.

TECHNICAL INFORMATION

Design of Concrete Mixtures

Two principal requirements of concrete are strength and durability — strength to perform the function of the structure and durability to resist exposure to the elements. These should be the governing considerations in the design of mixtures.

A third requirement is imposed by the fact that during placing, concrete must be workable. Still a fourth element in the design is economy. This is not in the same order of importance as strength and durability however. But a successfully designed concrete mix will achieve a proper balance between these four essentials.

Basis for design of mixtures. The design of concrete mixtures is based principally upon the relationship between the properties of the cement and the amount of mixing water used. A fundamental law — the *water-cement ratio* — has been established which may be stated as follows: For mixtures using sound and clean aggregates (sand and gravel), *the strength and other desirable properties of concrete under given job conditions are governed by the proportion of mixing water to the amount of cement.*

This principle may be more readily understood if the cement and water are thought of as forming a paste which, on hardening, binds the aggregate particles together. The strength of this paste is determined by the proportions of cement and water. Increasing the water content dilutes the paste and reduces the strength and durability of the concrete. A thin or watery paste, therefore, weakens binding power and does not form a strong or durable concrete.

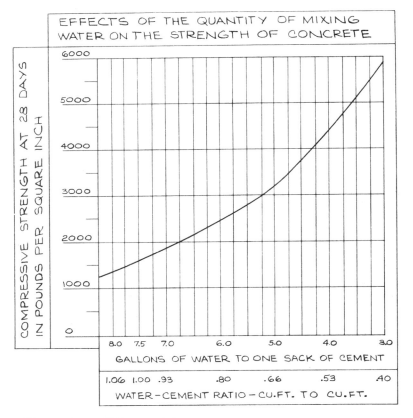

Figure 9-1. *Effects of Water on Strength of Concrete*

The hardening of the cement-water paste results from the chemical reactions between the water and cement, called *hydration*. During this process, a certain amount of water chemically combines with the cement to become a part of the permanent solid structure of the concrete and also causes heat to be developed. In order to be complete, these reactions require time, the presence of moisture, and favorable temperatures. This total process is known as the *curing* of concrete.

To obtain plastic workable mixtures, more water is used than can be permanently (chemically) combined with the cement. There is a reason for this. During the curing process, if it is controlled correctly, the excess water will evaporate because of the generated heat, the

sun, and air temperatures. Therefore, care should be taken that too much water is not lost, thereby stopping complete hydration. In order to counteract such excess loss of water, certain methods are used such as wetting down the concrete surfaces with water, covering the surface with straw, sand, or burlap, or covering with certain types of paper to keep water from evaporating too quickly. Thus, both factors — the quantities of water used and the extent and process of curing — directly affect the binding power (complete hydration) of the paste, and through it, the strength and durability of the concrete.

In view of the dependence of the properties of the concrete upon the quality of the paste, it will be seen that, as far as proportioning is concerned, the relative amounts of aggregate and cement are important only insofar as they affect workability, provided that the mixes are truly plastic and workable. Designing a concrete mix, therefore, consists of selecting the water-cement ratio that will produce concrete with the desired resistance to exposure and with the required strength, and finding the most suitable combination of aggregates that will give the necessary workability when mixed with cement and water in this ratio. Figure 9-1 illustrates the compressive strength of concrete in lbs. per sq. in. (PSI) when the indicated water-cement ratios are used.

Selection of Consistency

The design of a concrete mix involves consideration of the requirements to be met in handling and placing the concrete, and the factors affecting economy. In describing the character of fresh concrete, three words are most often used: consistency, plasticity, and workability.

"Consistency" is a general term that describes the state of the fluidity of the mix: it includes the entire range of fluidity from the driest to the wettest possible mixtures. The word "plasticity" is used to describe a consistency of concrete which can be readily molded but which permits the fresh concrete to change form slowly if the mold is removed. "Workability" describes the specific concrete mix. For example, a fairly dry mix with larger aggregates can be used for a footing, and a more fluid mix can be used for a thin wall with reinforcings, yet each concrete mix has the correct workability to meet the requirements of its end use in the area in which it is to be placed in the construction of a building.

The slump test. This test measures the consistency of a concrete mix. It may be performed in the laboratory or in the field. The test specimen is formed in a mold made of #16-gauge galvanized metal, shaped like a cone, open at the top and bottom. The base of the cone is 8 in. in diameter, the top opening is 4 in. in diameter, and the height of the cone is 12 in. The top and bottom openings must be parallel and at right angles to the center line of the cone. The mold must be provided with foot pieces and handles as shown in Figure 9-2.

How the test must be performed. Samples of concrete for the test specimen should be taken at the mixer or, in the case of ready-mixed concrete, from the transportation vehicle during discharge. The sample of concrete from which test specimens are made must be representative of the entire batch. Such samples are obtained by repeatedly passing a scoop or pail through the discharging stream of concrete, starting the sampling operation at the beginning of discharge and repeating it until the entire batch is discharged. The sample thus obtained is transported to the place of molding of the specimen and, to counteract segregation, the concrete must be mixed with a shovel until it is uniform in appearance. The location in the construction of

the batch of concrete thus sampled must be noted for future reference. As for the test itself, the mold is dampened and placed on a flat,

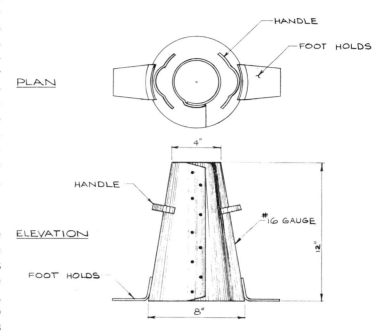

Figure 9-2. Mold for Slump Test

moist, nonabsorbent surface. From the sample of concrete obtained as described, the mold is filled in three layers, each approximately one-third the volume of the mold. Each layer is then rodded with 25 strokes of a 5/8-in. rod, 24 in. in length, bullet-pointed at the lower end. The strokes must be distributed in a uniform manner over the cross section of the mold and must penetrate into the underlying layer. The bottom layer is rodded throughout its depth. After the top layer has been rodded, the surface of the concrete is struck off with a trowel so that the mold is exactly filled; the mold is immediately removed from the concrete by being raised carefully in a vertical direction. The slump is measured immediately by determining the difference between the height of the mold and the height at the vertical axis of the specimen (see Figure 9-3).

The consistency is then recorded in terms of inches of subsidence of the specimen during the test, these values being known as the *slump*. After the slump measurement is completed, the side of the concrete

mix will crumble, segregate, and fall apart as at B in Figure 9-3. Recommended slumps for various types of structures are shown in Table 9-1.

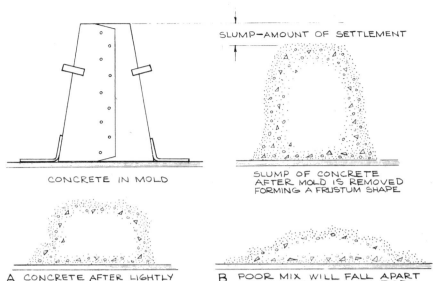

SLUMP—AMOUNT OF SETTLEMENT

CONCRETE IN MOLD

SLUMP OF CONCRETE AFTER MOLD IS REMOVED FORMING A FRUSTUM SHAPE

A CONCRETE AFTER LIGHTLY TAPPING SIDE WITH ROD. MIX WILL HOLD TOGETHER

B POOR MIX WILL FALL APART AFTER TAPPING WITH ROD

Figure 9-3. Slump Test

Table 9-1. *Recommended Slumps for Concrete*

Type of Structure	Slump in Inches	
	Maximum	Minimum
Massive sections; pavements and floors laid on ground	4	1
Heavy slabs; beams or walls	6	3
Thin walls and columns; ordinary slabs or beams	8	4

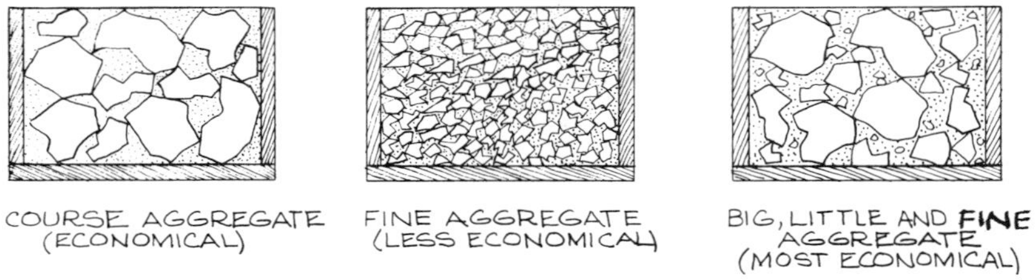

COURSE AGGREGATE (ECONOMICAL)

FINE AGGREGATE (LESS ECONOMICAL)

BIG, LITTLE AND FINE AGGREGATE (MOST ECONOMICAL)

Figure 9-4. Aggregate Combinations

frustum should be tapped gently with the tamping rod. The behavior of the concrete under this treatment is a valuable indication of the cohesiveness, workability, and placeability of the mix. A well-proportioned, workable mix will gradually slump to lower elevations as at A in Figure 9-3, and retain its original identity, whereas a poor

Selection of Aggregate Combinations

In determining the proportions of materials in a concrete mix, it is desirable to arrive at those proportions which will give the most economical results consistent with the proper placement. The relative proportions of fine and coarse aggregates and the total amount of aggregate that can be used with fixed amounts of cement and water will depend not only on the consistency of concrete required but also on the grading of each aggregate. A combination of aggregates made up largely of coarse particles presents less total surface to be coated with cement paste than does an aggregate consisting of fine particles and is therefore more economical. Consequently, for purposes of economy, it is desirable to use the lowest proportion of fine aggregate that will properly fill the "void" spaces in the coarse aggregates. Aggregates that are graded so that they contain many sizes are more economical than aggregates in which one or two sizes predominate, because the former contain fewer voids. The small particles fill the spaces between the larger particles, which otherwise must be filled with cement paste (see Figure 9-4). A properly proportioned combination of well-graded fine and coarse aggregate contains all sizes between the smallest and largest without an excessive amount of any one size.

Selection of Materials

Cement. All cement should conform to the standard specifications and test of the American Society for Testing Materials (ASTM).

Water. The water should be suitable for drinking unless it is known by test or experience that other waters are satisfactory. Sea water that contains salt should not be used for mixing concrete because it will corrode the steel reinforcements and chemically weaken the paste.

Aggregate. This word refers to an inert mineral filler used with the water-cement paste. All aggregates should consist of inert materials that are clean, hard, strong, and durable. The particles should not be covered with clay or dirt because this prevents proper bonding of the paste; it is important that only clean aggregates are used in concrete. The size and grading of aggregates are important from the standpoint of economy, because they have a very decided effect on the workability of the concrete and the amount of concrete that can be produced from a given volume of cement when used with a fixed proportion of water.

The maximum size of aggregate is governed by the nature of the work. In thin slabs or walls, the largest pieces of aggregate should not exceed about one-fifth to one-fourth the thickness of the section of concrete being placed (see Figure 9-5).

In reinforced concrete, specifications usually limit the maximum size to three-fourths of the minimum clear spacing between reinforcing bars.

Bank-run gravel. Sand and gravel suitable for concrete often are found as bank-run gravel — the natural mixture of sand, gravel, and stone as it comes from a gravel bank. Because bank-run gravel is rarely found in the proper mixture of sizes, it may therefore be necessary to screen it into various sizes (see Figure 9-6).

Gravel not larger than 1 1/2 in. in diameter is suitable for most construction. Where big stones occur in large numbers in bank-run gravel, it is necessary to screen the material first over a 1 1/2-in. screen. Material failing to pass this

screen is discarded. Material passing through the 1 1/2-in. screen (8 openings to the foot) should then be passed over a No. 4 screen (4 openings to the inch). That part

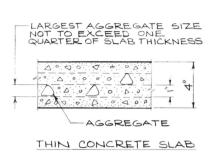

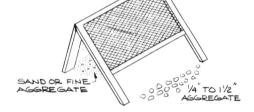

Figure 9-5. Aggregate Size

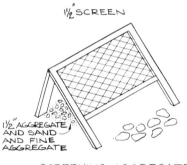

Figure 9-6. Screening Aggregate

passing through the No. 4 screen is sand and fine aggregate, while the material remaining is 1/4 in. to 1 1/2 in. aggregate which is suitable for most constructions.

Lightweight concrete. Lightweight aggregate characterizes this type of concrete, which is used as fill because it is light in weight, has insulating value, and is nailable.

Mixing Concrete

Concrete may be mixed by the following three methods:

1. Mixing completely in a stationary mixer, and transporting the mixed concrete to the point of delivery in a truck

agitator or in a truck mixer operating at agitator speed in which the mix is agitated until its final deposit into the forms. This is known as ready-mixed concrete.

2. Mixing partially in a stationary mixer and completing the mixing in a truck mixer. This is called "shrink-mixed" concrete.

3. Dry-mixing (dry-batching) all the materials at the central concrete plant and continuing the mixing en route to the job in a truck mixer. (Water is added at the site before the concrete is deposited into the forms.) This is called transit-mixed concrete.

The modern transport mixing trucks are provided with water tanks and measuring devices; mixer capacities range from 1 to 10 cu. yd. maximum capacity.

The amount of mixing can increase or decrease the strength of the concrete. The greatest gain in strength is during the mixing time from zero to two minutes; from six to ten minutes, there is little increase in the strength of the concrete. Figure 9-7 indicates how the compressive strength of concrete can be affected through the mixing time, as well as how the concrete strength increases after seven and twenty days, after three months, and after one year.

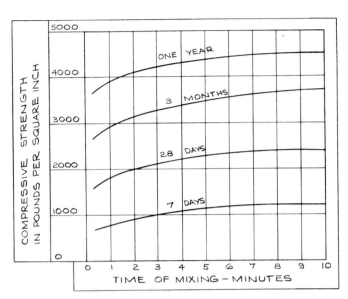

Figure 9-7. *Effect of Mixing Time on Compressive Strength*

REVIEW EXAMINATION

1. What are the two principal requirements of concrete?

2. The strength of concrete is governed by what?

3. In describing the character of concrete, what three terms are most often used?

4. What does the slump test measure in a concrete mix?

5. In thin concrete slabs or walls, the largest pieces of aggregates should not exceed what fraction of the thickness of the section?

6. Concrete mixed in 4 min. will achieve a greater strength than a concrete mixed in 8 min. — true or false?

7. A concrete mix of 5 1/4 gal. of water to one sack of cement will produce a compressive strength of how many lbs. per sq. in. after 28 days?

ASSIGNMENT

1. Prepare a plan and elevation scale drawing of the mold used for the slump test. Dimension the mold and label all parts.

2. In fifty words or less, describe how the slump test must be performed.

*Laboratory Report of a Concrete
Design Mix*

Reproduced here is a typical
concrete design mix laboratory report by the director of a bureau of
building design to the construction
company assigned to do the job.

In the laboratory report for the
1500 PSI lightweight fill, 500 lbs.
of portland cement are used in 1
cu. yd. of concrete mix. Because a
bag of cement weighs 94 lbs. it is
obvious that 500/94 = 5.32 bags or
cu. ft. of cement. To find the
water-cement ratio, divide the gal-

PO-105 New 30th Precinct
Station House
Contract No. 1 Gen. Const.
Concrete Design Mix

Gentlemen:

This is to acknowledge receipt of your letter of
Feb. 23, 197- together with four (4) copies of laboratory reports No. 3534 dated Jan. 15 and Jan. 18, 197-
and the water-cement ratio curves as prepared by
Certified Laboratories, Inc.

This data is for the 1500 PSI lightweight concrete
fill, 2500 PSI average concrete, 2500 PSI + 25% controlled, 2500 PSI + 25% pump concrete to be supplied by
the concrete contractor [name of contractor] for use at
the above project and will be filed with the Building
Department as required by the Building Code.

	1500 PSI Ltw. Fill	2500 PSI Ave.	2500 PSI Controlled	2500 PSI Pump
One cu. yd. dryweights portland cement	500 lbs.	564 lbs.	495 lbs.	591 lbs.
Concrete sand	625 "	1205 "	1275 "	1445 "
Coarse aggregate	725 " nytrolite	1850 " (3/4" stone)	1850 " (3/4" stone)	1550 "
Water (gallons)	40.9	37.1	36.9	38.9
A.E. Agent (darex)	5.3 oz.	4.5 oz.	5.3 oz.	5.3 oz.
W/C ratio	7.7	6.18	7.01	6.19
Weight per cu. ft.	97.8	145.5	145.5	144.4
Lightweight fines	450 lbs.	-	-	-
Slump	5" ± 1"	5" ± 1"	5" ± 1"	5" ± 1"

Very truly yours,
Director
Bureau of Building Design

Table 9-2. *Net Water-Cement Ratios for Concrete*

Type (Class) or Location of Concrete Structure and Degree of Exposure	Water-Cement Ratio by Weight	
	Severe climate, Wide Range of Temperature, Long Periods of Freezing or Frequent Freezing and Thawing	*Mild Climate, Rainy or Arid, Rarely Snow or Frost*
A. Concrete in portions of structures subject to exposure of extreme severity such as exposed foundation walls, piers and parapets, copings, and concrete in the range of fluctuating water levels and spray	0.45 ± 0.02	0.55 ± 0.02
B. Concrete in exposed structures where exposure is less severe than in A, such as the exterior of mass concrete and other exposed concrete not covered by A	0.50 ± 0.02	0.55 ± 0.02
C. Concrete in structures to be covered with backfill, or to be continually submerged or otherwise protected from the weather, such as foundations or substructures	0.58 ± 0.02	0.58 ± 0.02
D. Concrete subject to attack by sulfate alkalis in soil and ground waters, but to be placed during moderate weather	- - - - -	0.50 ± 0.02
E. For concrete same as D but to be placed during freezing weather, when calcium chloride would normally be used in mix	0.45 ± 0.02	- - - - -
F. Concrete deposited by tremie in water	0.45 ± 0.02	0.45 ± 0.02

Concrete Manual, U.S. Department of the Interior, Bureau of Reclamation, 7th ed.

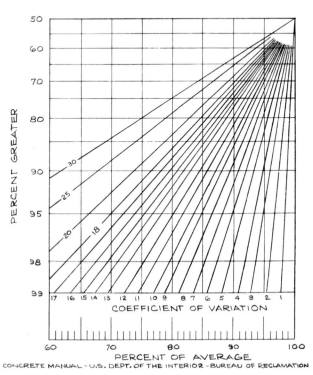

Figure 9-8. *Cumulative Probability Curves for Different Coefficients of Variation*

lons of water by the cu. ft. of cement, or

$$\frac{40.9}{5.32} = 7.7 \text{ W/C ratio}$$

Explanation and interpretation of laboratory report. The computation of proportions for concrete mixes can best be explained by means of a specific example.

Assume that a concrete foundation is considered which will not be exposed to freezing and thawing. This condition will permit the use of class (type) C concrete with a water-cement ratio of 0.58 (see Table 9-2).

The designers have specified a design strength of 2500 PSI at 28 days. Criteria generally accepted by Bureau designers require that the strength of 80% of the test specimens be greater than the design strength designated on the drawings. Therefore, an average strength of 2870 PSI will be required in order for 80% of the tests to fall above 2500 PSI.

Figure 9-8 shows cumulative probability curves for different coefficients of variation. This chart provides a simple means of determining the required average strength for any particular project in order for 80% of the tests to fall above a specified design strength. If it is assumed that concrete for a particular project will have a coefficient of variation of 15%, the slanting line for the coefficient of 15% is found to intersect a horizontal line extending from "80% greater" at a point directly above 87% of the average strength. Thus, for a coefficient of variation of 15%, 80% of the test specimen will have strengths higher than 87% of the average strength.

A common stipulation is that 80% of the test specimen shall produce strengths greater than 3000 pounds per sq. in. Then the required average strength is 3000/0.87 = 3448 PSI in order that 80% of the tests fall above 3000 pounds per sq. in.

In the computation of proportions, the following design criteria and mix materials will be assumed:

1. Type II non-air-entraining cement with a specific gravity of 3.15. A suitable pozzolan is available at a cost lower than cement, and laboratory investigations have indicated some danger of alkali-aggregate reaction. The cementing materials will therefore be composed of 30% pozzolan and 70% portland cement.

2. Coarse aggregate with a specific gravity of 2.68.

3. Sand with a specific gravity of 2.63 and a fineness modulus of 2.75.

4. A 3-in. slump and a maximum size aggregate of 3 in. will be satisfactory.

See Table 9-3.

Special Types of Concrete

Admixtures. Admixtures may be divided into two categories: those for mixing into concrete, and those for surface application or finish.

Concrete admixtures used for mixing into concrete include accelerators, retarders, finely divided powders, plasticizing agents, air-entraining agents, waterproofing compounds, and color pigments.

Surface applications or finishes for concrete consist of hardeners, color pigments, special aggregates, sealers, abrasive materials, and fillers and patchers.

Air-entrained concrete. The term describes a type of concrete which

Example of the computation of a trial mix.*

Mix Ingredients	Weight (in pounds per cu. yd.)	Conversion of Weight to Volume	Solid Volume (cu. ft. per cu. yd.)
Water (see Table 9-3)	204	$\dfrac{204}{62.3}$	= 3.27
Cement and Pozzolan W/C for durability, class C (see Table 9-2) = 0.58 W/C for strength = 0.58 $\text{cementing materials} = \dfrac{\text{water content}}{W/(C+P)}$ $= \dfrac{204}{0.58} = 352 \text{ pounds}$			
portland cement $= 352 \times 0.70 =$	246	$\dfrac{246}{3.15 \times 62.3}$	= 1.25
pozzolan $= 352 \times 0.30 =$	106	$\dfrac{106}{2.50 \times 62.3}$	= 0.68
Air (see Table 9-3) = 3.5 percent $0.035 \times 27 =$			= 0.95
All ingredients except aggregates =	556		6.15
Aggregate Volume = 27–6.15 = percent sand (see Table 9-3) = 28 percent			20.85
Volume of sand $= 28 \times 20.85 =$			5.84
Volume of coarse aggregate = 20.85–5.84 =			15.01
Weight of sand = $5.84 \times 2.63 \times 62.3 =$	957		
Weight of coarse aggregate = $15.01 \times 2.68 \times 62.3 =$	2507		
Total	4,020		27.00

*From Concrete Manual, 7th ed.

Table 9-3. *Approximate Air and Water Contents Per Cubic Yard of Concrete and the Proportions of Fine and Coarse Aggregate*

For Concrete Containing Natural Sand with an FM of 2.75 and Average Coarse Aggregate, and Having a Slump of 3 to 4 in. at the Mixer

Air-Entrained Concrete			Non-Air-Entrained Concrete				
Sand percent of total aggregate by solid volume	Average water content. lbs./cu. yd.	Approximate percent entrained air	Sand percent of total aggregate by solid volume	Average water content lbs./cu. yd.	Recommended air content, percent	Percent dry-rodded unit weight of coarse aggregate per unit volume of concrete	Max. size of coarse aggregate, inches
3/8	41	8	322	58	3.0	352	61
1/2	52	7	306	50	2.5	336	53
3/4	62	6	283	42	2.0	316	45
1	67	5	267	37	1.5	300	41
1-1/2	73	4.5	245	33	1.0	280	36
2	76	4	229	30	0.5	266	33
3	81	3.5	204	28	0.3	242	31
6	87	3	164	24	0.2	210	28

From *Concrete Manual*, 7th ed.

results when air-entraining agents are added either in the manufacturing of the cement or during the mixing of the concrete to cause millions of tiny air bubbles to develop within the entire mix.

This type of concrete was developed for road-building to counteract freezing and to withstand the salt used to melt ice. Today air-entrained concrete is used extensively in building construction because it has fine workability and cohesiveness and also because the entrained air keeps the aggregates from separating and the water from rising to the top of the concrete.

Concrete for small construction. Table 9-4 summarizes mixtures suitable for small projects. Ready-mixed concrete for small work, packaged in paper bags, is available from building materials suppliers.

Table 9-4. *Suitable Mixtures for Various Small Concrete Construction Projects*

Type of Construction	Cement by Volume	Sand by Volume	Aggregate by Volume
Foundations and footings	1	3	5
Driveways, sidewalks, steps	1	2-1/4	3
Small floor slabs laid on earth	1	2-1/2	3

10

Placing and Pouring Concrete

INTRODUCTION

In this unit are discussed the various kinds of forms used in concrete work, the care and supervision that must be exercised in placing concrete into forms to prevent "segregation" and "laitance," the precautions that must be taken during the curing of concrete, and the problems of expansion and contraction, including the provisions of construction joints to prevent the cracking of concrete work.

TECHNICAL INFORMATION

Forms for Concrete

The wood or metal construction that holds the concrete in place until it has hardened is called a form. Forms must be carefully put together to conform to the shape and dimensions of the concrete as shown on the plans, and are required to be substantial and sufficiently tight to prevent the leakage of water from the wet concrete. Consideration should be given to the cost of the formwork, because it constitutes a large part of the expense of concrete construction. Forms should be constructed with economy and simplicity whenever possible, with special attention given to ease of erection and stripping down after the concrete has

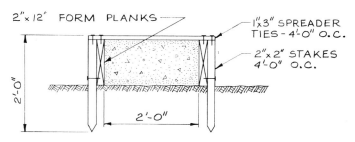

Figure 10-1. Form for Footing

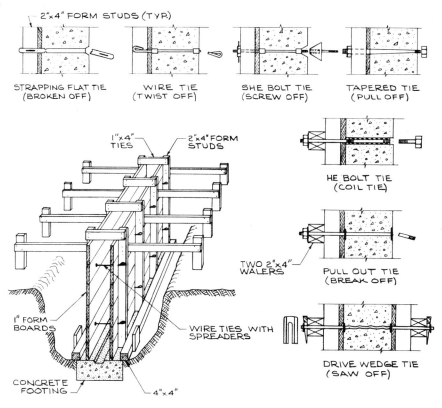

Figure 10-2. Tongue and Groove Form Boards and Ties

75

hardened. Carefully dismantled forms can be re-used numerous times.

Typical wood forms. Spruce and pine lumber are well-suited for wood forms, because they do not stain the exposed concrete surface. Green or partially seasoned lumber is preferable, because dry lumber swells when wet; but whether the wood form lumber is dry or seasoned, it must be thoroughly wetted before the concrete is poured. Frequently wood forms are oiled or treated to fill the pores of the wood to prevent the absorption of water

from the concrete and further provide for a smoother surface, thus preventing the concrete from sticking to the forms when they are removed.

The typical wood form for a footing (see Figure 10-1) may be constructed with 2 × 12-in. planks held in place by 2 × 2-in. stakes, 2 ft. long, driven halfway into the ground. The stakes are held together by 1 × 3-in. spacer ties, 4 ft. apart. Tongue-and-groove form boards, properly supported by ties, spreaders, stakes, and braces as shown in Figure 10-2, are typical in forms used for concrete foundation walls.

Major requirements of formwork. In building and installing forms, one important factor must be kept in mind — they must be made strong enough to support the weight of the wet concrete and to overcome any sideward forces or pressures exerted by the concrete. Concrete weighs approximately 150 lbs. per cu. ft. This mass can be better visualized by referring to Figure 10-3A, which shows a sectional view of a concrete foundation and the formwork necessary to hold the wet concrete. The wall is 6 ft. high and 1 ft. wide. If a 1-ft. length of this wall, as shown in Figure 10-3B, were to be considered, it would contain 6 cu. ft. of concrete, which has a total weight of 900 lbs. This weight causes a considerable sideward force against the forms, as indicated by A, B, C, and D. This force is overcome by proper bracing and by the use of form ties, which extend through the concrete and are fastened to the outer sides of the studs of the formwork.

Plywood forms. Plywoods have been developed that are suitable for concrete forms. The use of a special bonding agent, surface oiling, and edge sealing combine to make it a material that can be re-used many times. Large smooth panels (see Figure 10-4) minimize the number

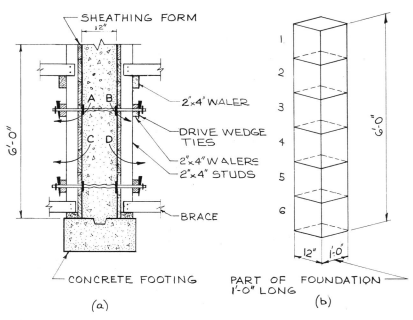

Figure 10-3. Cross Section of Concrete Foundation Showing Form Work

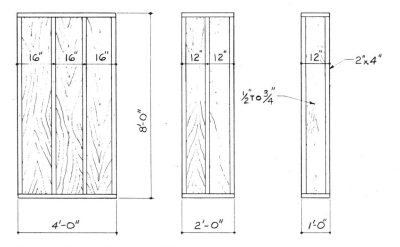

Figure 10-4. Plywood Forms

of joints and tend to simplify construction. The material is strong and rigid and therefore requires fewer supports. It is identified by a special grade mark and is available in sizes up to 4 × 12 ft. and in thicknesses ranging from 1/2 to 3/4 in. It makes a smooth surface on the concrete and can be bent to form curved surfaces. Further advantages are that it can be sawed easily and nailed close to the edge without previous drilling and without splitting.

Formwork around beams. Steel beams that are to be encased in concrete are known as fireproofed steel beams. Figure 10-5 illustrates two methods of constructing formwork of this type that will receive the wet concrete. Snap-tie hangers in U-shaped form hold the tongue-and-groove beam sides and bottoms, studs, stringers, and joists. After the concrete is poured and hardened, the forms can easily be removed. Plywood forms for concrete columns, as shown in Figure 10-6A, are held together with special metal locking squares which can be quickly installed and dismantled. For concrete slabs and reinforced concrete beams, as shown in Figure 10-6B, 1-in. tongue-and-groove pine formboards or plywood are used for beam sides, and 2-in. stock or thick plywood for the beam bottoms.

Metal forms. Steel and cast iron make excellent forms for concrete and are used extensively where forms can be re-used many times. These may be units such as wall panels, which can be assembled in a variety of shapes or special forms for units such as blocks, ornamental forms, etc. Because such metal forms are practically indestructible, the cost per unit may become quite small as the usage factor increases. Steel forms for making regular spaced concrete joists are shown in Figure 10-7. After the slab and con-

crete joists become self-supporting, the forms are removed. Steel domes may be used for two-way ribbed (waffle-type) slab construction (see

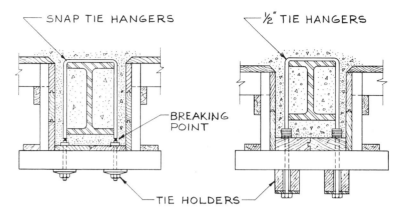

Figure 10-5. Formwork for Beam Concrete Fireproofing

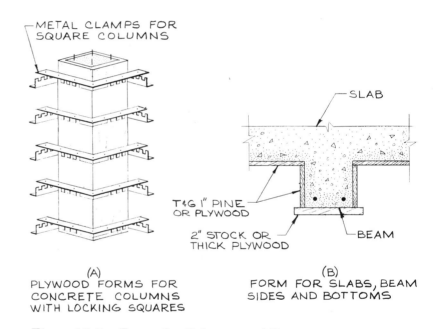

(A)
PLYWOOD FORMS FOR
CONCRETE COLUMNS
WITH LOCKING SQUARES

(B)
FORM FOR SLABS, BEAM
SIDES AND BOTTOMS

Figure 10-6. Forms for Columns and Beams

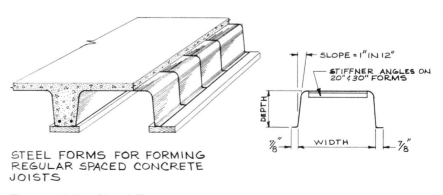

STEEL FORMS FOR FORMING
REGULAR SPACED CONCRETE
JOISTS

Figure 10-7. Metal Forms

Figure 10-8). They are rigid deep-drawn, one-piece units which can easily be removed when the concrete has hardened. Permanent metal forms such as those shown in Figure 10-9 serve as the structural subfloor for concrete roofs and floor slabs.

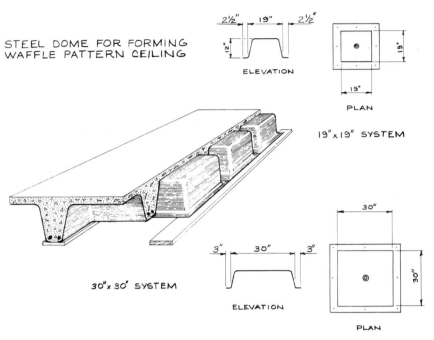

Figure 10-8.   Steel Domes

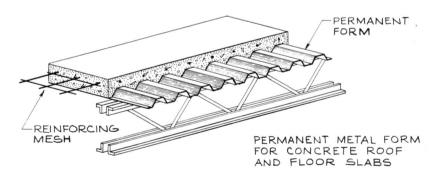

Figure 10-9.   Rib Form

Synthetic forms.   Glass fiber, reinforced polyester plastic and paper pulp are used as formwork for roof and floor panels, for domes, and for two-way ribbed (waffle-type) slab construction in conjunction with flat-plate design. These forms are light in weight, flexible, and easily removed and rehandled. They require no oiling and are not subject to corrosion.

Removal of Forms

Forms should be removed in such a manner as to insure the complete safety of the structure. Where the structure as a whole is supported on shores, the removable floor forms, beams and girder sides, column and similar vertical forms may be removed, provided the concrete has achieved sufficient strength. In no case should the supporting forms or shoring be removed until the concrete members have acquired sufficient strength to support safely their weight and the loads for which they were designed.

Placing Concrete

No element in the entire cycle of concrete production requires more care than the final operation of placing concrete at the ultimate point of deposit.

Before placing concrete, all débris and foreign matter should be removed from places to be occupied by the concrete. The forms that receive the concrete should be thoroughly wetted or oiled or otherwise treated as required. Temporary openings should be provided where necessary to facilitate cleaning and inspection immediately before depositing concrete; these should be placed in such a way that excess water used in flushing the forms may be drained out.

Prevention of segregation. With a well-designed mixture delivered with proper consistency and without segregation, placing of concrete is greatly simplified, but even in this case care must be exercised to continue to prevent segregation and to see that the material flows properly into the corners and angles of forms and around the reinforcements. Constant supervision is essential to insure such complete filling of the form and to prevent the rather common mistake of depositing continuously at one point (see Figure 10-10) and allowing the material to flow to distant points. Flowing over long distances will cause segregation, especially the separation of water and fine particles from the rest of the mass. Excessive amounts of tamping or shoveling in the forms will also cause the materials to separate. When the concrete is properly proportioned, the mass will be thoroughly consolidated with very little shoveling. Light spading of the concrete next to the forms will prevent honeycombing and will be conducive to better surface finishes. In placing concrete in deep layers, a gradual increase in the water content of the upper portion is quite certain to result from the increased pressure on the lower portions. The excess water should be worked to a low point, without actually causing flow, and be removed. It should be remembered that excess water in the upper layer is just as objectionable as excess water in a mix.

Laitance. The whitish, chalklike substance of very little strength that forms on the upper surface of the concrete is called *laitance*; this laitance layer can be seen at the top of each day's work. Two preventive measures are necessary: (1) the concrete mixture should be of such consistency that laitance will be minimized; and (2) before each day's pouring, the laitance must be washed off and completely removed so that proper bonding between the pours is accomplished. If the laitance is not removed, bonding is not accomplished, the structure is weakened, and water can penetrate through this weakened joint (see Figure 10-11).

Curing of Concrete

The chemical reactions that take place as cement hardens continue for a very long period and result in a gradual increase in the strength of

the concrete. In order for these reactions which constitute "curing" to continue, moisture must be present and the temperature must

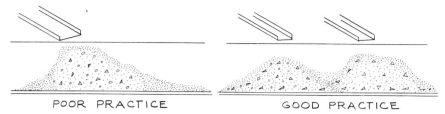

POOR PRACTICE   GOOD PRACTICE

DEPOSIT CONCRETE AT VARIOUS POINTS IN ORDER TO PREVENT SEGREGATION

Figure 10-10. Depositing Concrete into Forms

be favorable. A layer of water is always present on the surface when the concrete is placed, but if the evaporation of the water is not controlled and too much evaporation occurs, the chemical reactions required to achieve the full strength of the concrete will not be accomplished. Water evaporation should be controlled for five days or longer in one of the following ways:

1. By wetting down the concrete surface periodically with water;
2. By covering the surface with straw, sand, or burlap which is kept moist;
3. By covering the surfaces with certain types of paper to keep the water from evaporating.

When poured in winter concrete must be protected against freezing for at least 72 hrs. after being placed. The heat generated by the chemical reactions of the concrete is an important factor in maintaining the required temperatures during this period of time. If more heat is required, the mixing water and even the aggregate may be heated. Specifications often require that concrete be maintained above 70° F. for the first three days after placing, and above 50° F. for the first five days, but the materials should not be heated enough to raise the temperature of the concrete above 100° F.

After concrete has been placed, it can be protected by enclosing in

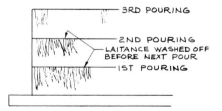

3RD POURING

2ND POURING
LAITANCE WASHED OFF BEFORE NEXT POUR
1ST POURING

LAITANCE CAN ONLY FORM FROM THE PRESSENCE OF EXCESS WATER AT THE TOP OF POURED CONCRETE

Figure 10-11. Laitance Formed by Interrupted Pours

some manner and maintaining heat in the enclosed space. Covering a temporary framework with canvas tarpaulins or reinforced plastic sheets is a common method. The heat also may be supplied by steam which is allowed to escape into the enclosed space to provide moisture for curing, or by salamanders, which are crude coke-burning stoves simple to operate but difficult to control.

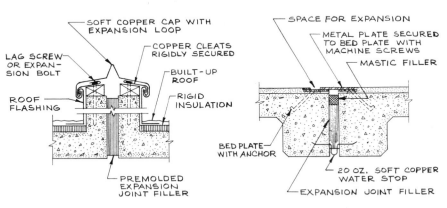

Figure 10-12. Expansion Joints through Floor and Roof

Shrinkage of Concrete

Concrete changes somewhat in volume with different conditions of exposure. Shrinkage takes place from the time of installation and continues for a period of several months upon exposure to dry air conditions.

The fundamental cause of shrinkage is the loss of water resulting from evaporation and from absorption by the aggregate particles, earth, or the forms. Shrinkage during the first stage of setting is quite rapid, the rate depending upon the rate of evaporation. In small thin sections of concrete exposed to rapid drying conditions, a great deal of shrinkage will result in cracking. This can be avoided by preventing the loss of water from the mass by some of the methods already described.

Volume changes are also caused by temperature changes: higher temperatures cause expansion and lower temperatures cause contraction.

Expansion Joints

Buildings over 300 ft. in length should be separated by a transverse joint providing a plane of separation so that free movement of the two adjacent parts may take place. Reinforcement should never extend across the expansion joint, the break between the two sections should be complete, and exposed joints should be filled with an elastic joint filler. Figure 10-12 illustrates an expansion joint through a floor and one through a roof. Note the expansion joint filler and the provisions that are made in protecting the joint from water seepage.

How to calculate expansion for a 300-ft. concrete building. The coefficient of expansion for concrete is:

0.0000065 in. per change in 1° F., assuming a temperature difference from 0° in winter to 90° in summer where the building is to be located.

The formula is:

length (ft.) × 12 × coefficient of expansion × temperature difference

or   300 ft. × 12 × 0.0000065 × 90° = expansion in inches = 2.1 in.

Therefore the expansion joint at the center of the building is a minimum of 1.05 in.

REVIEW EXAMINATION

1. Name the four types of forms that can be used in concrete work.

2. Give the thickness of the wood form material used for concrete

beam sides; for concrete beam bottoms.

3. For plywood forms, what is the range of thickness of the plywood?

4. What type of form material may be used for two-way ribbed (waffle-type) slab construction?

5. In depositing concrete into the forms, name two ways in which segregation may occur.

6. What should be done with the laitance that forms on top of a concrete pour before the next pour can be made?

7. Name three methods in which water evaporation of freshly poured concrete can be reduced.

8. Freshly poured concrete must be protected against freezing for how long a period?

9. What is the fundamental cause of shrinkage in concrete?

10. A transverse expansion joint must be used in buildings that exceed a length of how many feet?

ASSIGNMENT

1. Sketch a waffle-type floor construction illustrating the form that is used.

2. Sketch a construction system illustrating the use of permanent forms.

3. Sketch a detail indicating a wood form used for a concrete footing.

SUPPLEMENTARY INFORMATION

Concrete Test Specimens

When concrete is placed, a sample of the concrete is taken to the testing laboratory to determine its compressive strength and to check that the requirements for which it is intended are met. This is done by depositing the concrete in 6 × 12-in. cast iron molds in general use by the testing laboratory. The depositing of the concrete into the mold must be done under strict supervision and according to specific, controlled conditions. The specimen should be marked, weighed, and stored. Standard cured test specimens should be tested for compressive strength as soon as practicable. Details of testing can be found in any comprehensive text on concrete.

Placing Concrete under Water by the Tremie Method

Concrete should not be deposited under water if it is possible to deposit in air, because there is always some uncertainty about the results obtained from placing it under water. The additional expense and delay of avoiding this method are nearly always warranted. Where conditions are such that concrete must be placed under water, expert supervision is necessary and every precaution must be taken to prevent the cement from being washed out of the mix. A fairly rich mix, containing at least seven sacks of cement per cu. yd. of concrete should be used.

One successful method of placing concrete under water is by use of a "tremie" (see Figure 10-13), a steel pipe or tube with funnel-shaped upper end long enough so that the lower end will reach the bottom of

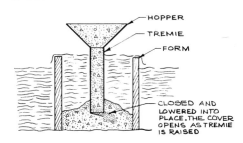

Figure 10-13. Tremie Method of Placing Concrete under Water

the work when the upper end is above the water level. When placing operations are to start, the lower end of the tremie has a special operable closure cover to prevent the entrance of water or the escape of concrete. The tremie is filled with concrete and lowered into position. When the lower end reaches the bottom of the work, the cover opens and the concrete is allowed to flow out slowly. As concreting proceeds, the lower end is kept submerged in the concrete for several feet and the tremie kept constantly full of concrete.

Vibrating Concrete

Vibrators are used when concrete has been poured into the form.

They insure that the concrete completely surrounds the reinforcing, they eliminate air pockets and, most important, they permit the mixes to be made with less water. The time allowed for vibrating and the spacing of the insertion points for the vibrator(s) must be carefully controlled because too much vibration causes a separation of the aggregates, with the larger aggregates falling to the bottom of the form.

Pumping Concrete

The transporting of concrete from the area of mixing (be it job mix or transit mix) to the area where it is to be poured has been a difficult, time-consuming, and costly operation, particularly for multistoried buildings. Methods have now been developed whereby concrete can be pumped directly from place of delivery to areas of placement.

When the pumping method has been selected for a job, then the mix, the cement-water ratio, the aggregate composition, and the height to be pumped all have to be tested and found satisfactory before any concrete can actually be installed.

11

Building with Concrete

INTRODUCTION

This unit reviews the major uses of concrete in the building industry. Reinforced concrete floor and roof systems, concrete planks and decking, prestressed concrete, precast reinforced concrete panels used as curtain walls, thin shell construction, and concrete rigid frames are considered in detail.

TECHNICAL INFORMATION

Concrete is used in residential and small construction work for footings, foundations, slabs on grade, steps, walks, and innumerable other areas where the loads and forces imposed are great enough to require that the concrete be reinforced with steel rods or mesh. Concrete is called "reinforced concrete" when such steel rods or reinforcing mesh are installed.

Reinforced Concrete

When steel and concrete are combined, the compressive strength of concrete and the tensile strength of steel are used to their best advan-tage. Concrete can develop a compressive strength of 6000 PSI, whereas steel can be manufactured to develop a tensile strength of over 150,000 PSI.

One of the many advantages of reinforced concrete is that supporting beam shapes can be controlled in size and contour, as shown in Figure 11-1.

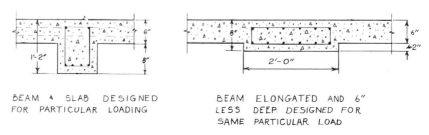

BEAM & SLAB DESIGNED FOR PARTICULAR LOADING

BEAM ELONGATED AND 6" LESS DEEP. DESIGNED FOR SAME PARTICULAR LOAD

Figure 11-1. Supporting Beam Shapes

There are four reinforced beam, column, and slab systems commonly used in building construction: 1-way, 2-way, flat slab 2-way, and flat slab 4-way. These are illustrated in Figure 11-2.

The *grid* or *pan floor system* is achieved by placing steel, plastic, or paper pans or domes on temporary forms and filling the voids with concrete and reinforcing steel. Steel, plastic, or paper pans are available in widths of 10 to 30 in., depths of 8 to 16 in., and in lengths

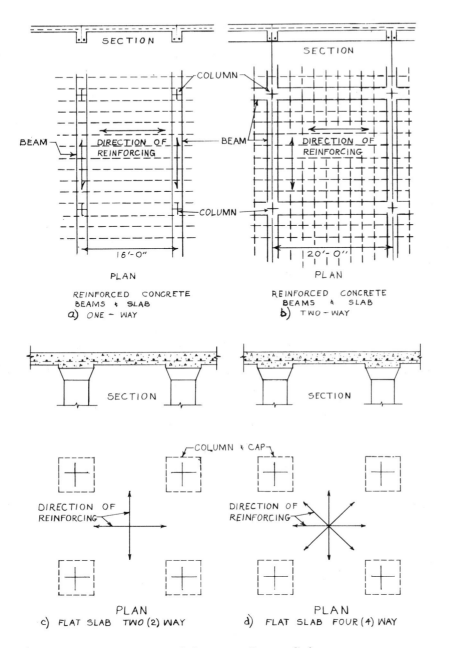

Figure 11-2. *Reinforced Concrete Beam, Column, and Slab Systems*

of 1'-0" to 3'-0", and with end closures tapered or straight for all the various pan sizes. Steel, plastic, or paper domes are available in sizes 18 × 18 in. to 30 × 30 in. with depths of 8 to 14 in. The width of the rib, controlled by the depth of

the pans or domes is generally 5 in. for the deeper type and 4 in. for the shallower type (see Figure 11-3).

The *tube-slab system* employs uniformly spaced metal or paper tubes as fillers, around which is poured a lightweight concrete with reinforcings (see Figure 11-4). The tubes are placed within a slab near the center so that the concrete can form around the tubes. The thicknesses of slab and the tube diameter and spacing vary as required for span loading conditions. Tube slabs range from 10 to 30 in. in thickness, with a span range from 18 to 80 ft. The tube-slab system is a practical system for many types of structural framing, because it results in a flat ceiling.

Concrete planks and decking can be defined as prefabricated reinforced concrete slabs or panel units that will span between supporting beams, walls, or partitions to form floors or roofs. Planks and decking are available in sizes ranging from 15 to 32 in. in width, 4 to 10 ft. in length, and 1 to 4 in. in thickness. Some such planks are made of lightweight aerated concrete, integrally cast into galvanized edgings of tongue-and-groove contour that form the tongue-and-groove edges on all four sides of the finished plank (see Figure 11-5). Special clips are provided to anchor the planks to the supporting members. Another type is the channel plank illustrated in Figure 11-6. It is 24 in. wide and 3-3/4 in. thick at its ends and 1 to 1-1/2 in. thick through its center.

The planks are fastened to the steel beams by means of metal clips. Where planks are used around stairwell openings or openings for ducts and shafts, the plank ends must be properly supported and correctly detailed.

Prestressed Concrete

When a reinforced concrete beam is designed, the steel reinforcing is calculated to overcome the bending within the beam. By prestressing the steel, the cross-sectional area of the beam can be reduced.

Prestressed concrete may be either pre-tensioned or post-tensioned. Pre-tensioning is accomplished by placing the reinforcing rods or wire cables in an empty form and pulling them to their required tensile stress by means of hydraulic jacks. The concrete is then poured into the form and allowed to cure. The jacks are released and the stress in the wires is transferred by the bond between the materials. That is, as the steel attempts to shorten and return to its original length, it compresses the beam by transferring the prestressing force by bond to the concrete.

In post-tensioning, the steel is stretched after the concrete has hardened. To achieve this, pipe conduits are placed in the concrete through which wire cables are stretched against the ends of the member. Final transfer of prestress from steel to concrete can be achieved by mechanical devices at the end anchorage.

Prestressed concrete is two to three times stronger than reinforced concrete. It allows for larger spans and the support of heavier loads.

Figure 11-7 illustrates two prestressed concrete beams of slightly different design. One is notched to receive a wood nailing strip or a dovetail anchor, while the other is provided with notched-out shelves for seating the prefabricated deck. Figure 11-8 shows how the decking is installed in each type of beam.

Double-tee floor and roof units have advantages insofar as they are precast, prestressed, and ready for installation, their light weight and long span resulting in economy of framing (see Figure 11-9). Here the double-tee unit can be cambered to compensate for deflection. The concrete forming the beam weighs about 105 lbs. per cu. ft. and attains a minimum strength of 3750

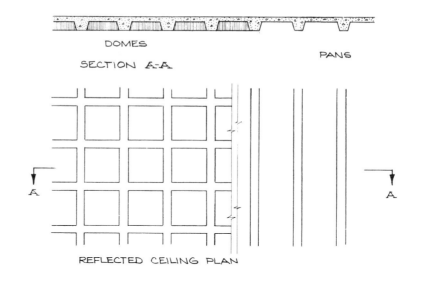

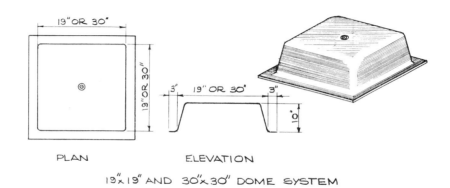

Figure 11-3. *Steel Domes and Pans*

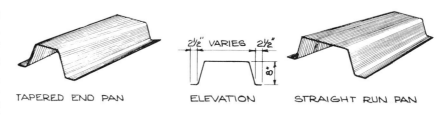

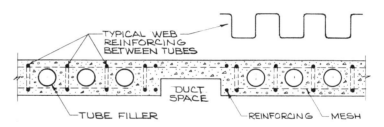

Figure 11-4. *The Tube-Slab System of Construction*

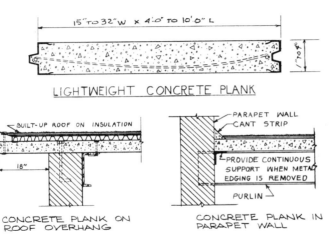

LIGHTWEIGHT CONCRETE PLANK

15" TO 32" W × 4'-0" TO 10'-0" L

BUILT-UP ROOF ON INSULATION

PARAPET WALL

CANT STRIP

PROVIDE CONTINUOUS SUPPORT WHEN METAL EDGING IS REMOVED

PURLIN

18"

CONCRETE PLANK ON ROOF OVERHANG

CONCRETE PLANK IN PARAPET WALL

CONCRETE PLANK ON FLAT ROOF

NO. 10 CLIP

NAIL HOLE ⅛" ⌀

1 ¾"

1 ¾"

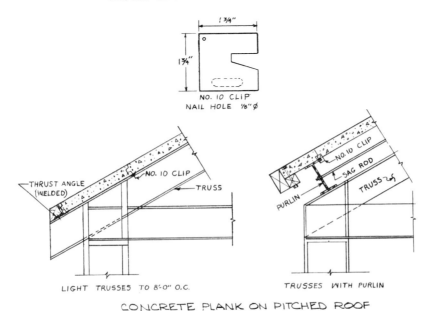

THRUST ANGLE (WELDED)

NO. 10 CLIP

TRUSS

NO. 10 CLIP

SAG ROD

PURLIN

TRUSS

LIGHT TRUSSES TO 8'-0" O.C.

TRUSSES WITH PURLIN

CONCRETE PLANK ON PITCHED ROOF

Figure 11-5. *Concrete Planks and Decking*

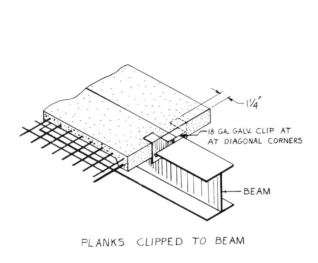

1 ¼"

18 GA. GALV. CLIP AT AT DIAGONAL CORNERS

BEAM

PLANKS CLIPPED TO BEAM

Figure 11-6. *Channel Planks*

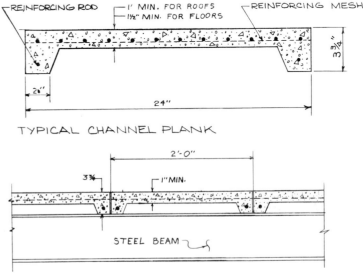

REINFORCING ROD

1" MIN. FOR ROOFS
1½" MIN. FOR FLOORS

REINFORCING MESH

3 ¾"

2 ½"

24"

TYPICAL CHANNEL PLANK

2'-0"

3 ¾"

1" MIN.

STEEL BEAM

CHANNEL PLANKS FOR FLOOR AND ROOFS

PSI after 28 days. The weight of the double-tee is 28 lbs. per sq. ft. The usual minimum bearing required is 4 in. on steel, and 6 in. on masonry. The top surfaces are purposely rough to increase the bond between the slab and the concrete topping. A typical single prestressed tee section is shown in Figure 11-10, and some typical construction details are shown in Figure 11-11.

Precast, prestressed, hollow-core sections are usually 4 or 8 ft. wide and may be 4, 6, 8, 10 and 12 in. in thickness. Spans range from 15 to 50 ft., capable of supporting superimposed loads of over 200 PSI. Figure 11-12 illustrates the core plank, together with some typical installation details. This type of floor may be used with reinforced concrete and masonry structures as well as with steel frame.

The cores may serve as a passage for utility lines and can also be utilized for warm air heating systems. The smooth undersurface of the slab formed by the planks may be painted directly when smooth or serve as an acoustical surface when textured.

Prefabricated panels of reinforced concrete are extensively used as curtain walls for buildings with steel or concrete framing. The panels of numerous designs may extend from floor to floor and are provided with quick anchoring devices that allow for ease of erection and installation.

A typical such panel is shown in Figure 11-13, including a partial elevation and a fastening detail of the panel. Figure 11-14 shows a variation of the panel over a window head and sill. A fastening detail is also shown.

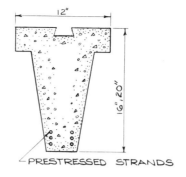

PRESTRESSED STRANDS

LOADS PER LINEAR FOOT FOR CONCRETE TEE JOISTS			
SPAN IN FT.	20	45	64
LB'S/LIN·FT. 16"	584	128	—
LB'S/LIN.FT. 20"	873	293	82

PRESTRESSED STRANDS

PRESTRESSED REINFORCED CONCRETE BEAMS- SPANS FROM 16'-0" TO 32'-0" IN INCREMENTS OF 2'-0"

Figure 11-7. Prestressed Concrete Beams for Floor and Roof Construction

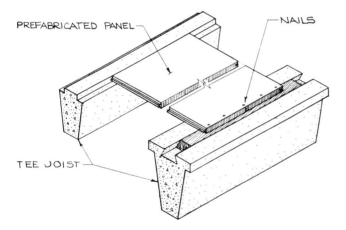

Figure 11-8. Application of Decking on Prestressed Beams

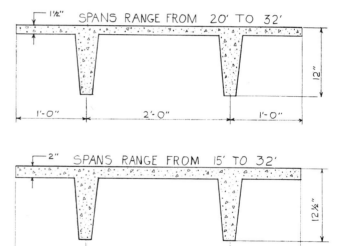

Figure 11-9. Double-Tee Floor and Roof Units

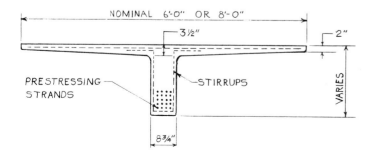

Figure 11-10. Typical Prestressed Tee Section

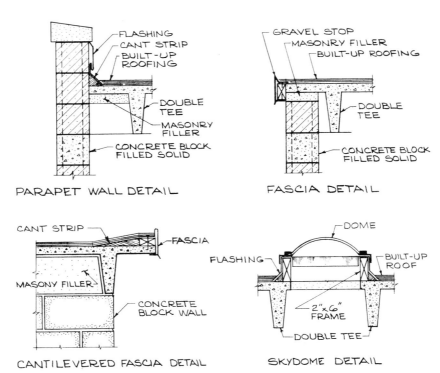

Figure 11-11. Typical Construction Details Using Double Tee Units

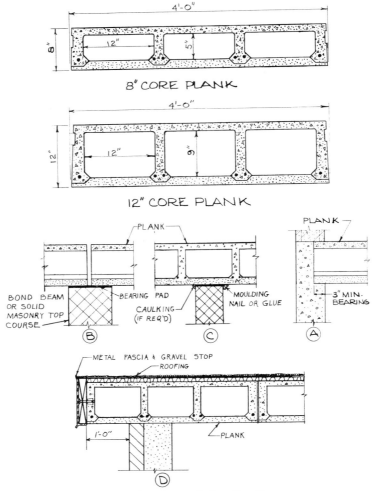

Figure 11-12. Core Planks and Applications

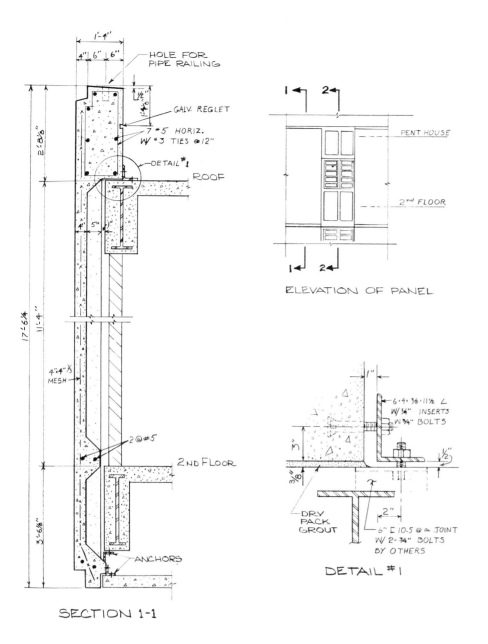

SECTION 1-1

ELEVATION OF PANEL

DETAIL #1

Figure 11-13. Concrete Panel as Curtain Wall

REVIEW EXAMINATION

1. Concrete is called "reinforced concrete" when either of what two materials are installed?

2. Concrete can develop a compressive strength of what PSI?

3. Steel can be manufactured to develop a tensile strength of over what PSI?

4. What are the four reinforced beam, column, and slab systems commonly used in building construction?

5. The tube-slab system employs what type of tubes as fillers, around which concrete is poured?

6. What are the various widths and

lengths available for concrete planks and decking?

7. Prestressed concrete is how many times stronger than reinforced concrete?

8. Spans for precast, prestressed, hollow-core sections may range from how many to how many ft.?

9. What are prefabricated panels of reinforced concrete extensively used as?

10. The double-tee unit is cambered to compensate for what?

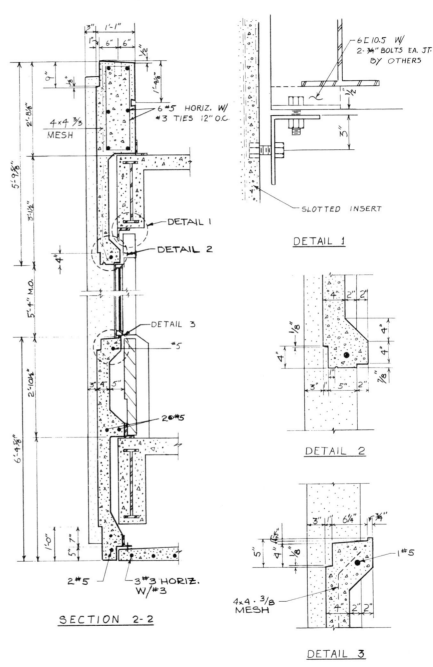

Figure 11-14. *Concrete Panel as Curtain Wall (cont.)*

1. By sketch or illustration show other applications of concrete in building.
2. Sketch a channel and a core

plank and give nominal dimensions.
3. Illustrate the difference between 1-way and 2-way ribbed construction.

SUPPLEMENTARY INFORMATION

Other Systems of Concrete Construction

Thin shell construction is a structural system in which concrete is used economically as a roof, wall, or a combination of both. The system permits a freedom of design shapes in both plan and section. It is strong and has a great capacity to carry unbalanced loads. Shells derive their strength from their ability to transfer loads by membrane stresses. These are direct stresses, such as compression, tension, and shear, acting over the entire thickness of the shell at any point. There is no bending of an element of the shell such as exists in a flat slab, except of minor magnitude caused by edge and end conditions. The thin-shell system of construction lends itself to many forms of building shapes, of which the barrel shell is most common. Other shapes, such as the multiple barrel shell, the "north light" shell, the butterfly shell, and the typical flat corrugated shell are some of the better-known types currently used (see Figure 11-15).

Placement of bars in shells may be in the form of reinforcing bars or a combination of bars and mesh. There are usually three layers of bars or mesh in a shell section (see Figure 11-16). The *diagonal bars* placed at the bottom of the shell resist diagonal tension, the *longitudinal bars* at the center resist the main tensile forces below the neutral axis, and the *transverse bars* at the top of the section resist some moment and shrinkage stresses.

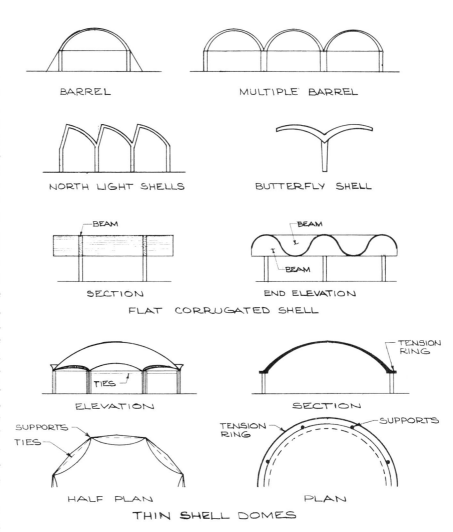

Figure 11-15. Concrete Thin Shell Structural Forms

Concrete rigid frames may be combined with shell construction, particularly in the long-type barrel form. The more common types of concrete rigid frames have the appearance of a rather flat arch and are most suitable for gymnasiums, auditoriums, hangars, markets, and

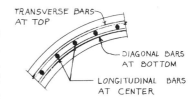

Figure 11-16. Placement of Reinforcing Bars in Thin Shell

armories (see Figure 11-17). Economical spans are generally between 60 and 100 ft., and the frames may be 20 ft. or more on center. As in most arches, the lower ends of the legs are usually connected with a tie or they are restrained from outward movement by buttresses. The frames may be provided with roof purlins, struts, or any of a number of precast concrete roof deckings.

In the *plank system*, precast concrete planks are easily fastened to steel, wood, or concrete beams. The longitudinal edges of the planks are tongue-and-grooved (see Fig-

ure 11-18) with bevels forming V-joints. Plank ends are square-cut and must be placed over beams.

The *bulb-tee system* (see Figure 11-19) is used for spanning between bulb-tee purlins. The sides of the tile (or plank) are chamfered for grout application and the ends may be tongue-and-grooved, or they may be square-cut.

The *dual-tee system* (see Figure 11-20) for roof decks incorporates the panel with a specially designed erosion-resistant sub-purlin of cold rolled 18-gauge steel, which is inserted on the job into a pre-cut kerf in the panel.

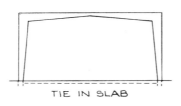

TIE IN SLAB

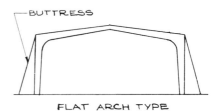

FLAT ARCH TYPE

Figure 11-17. Rigid Concrete Frames

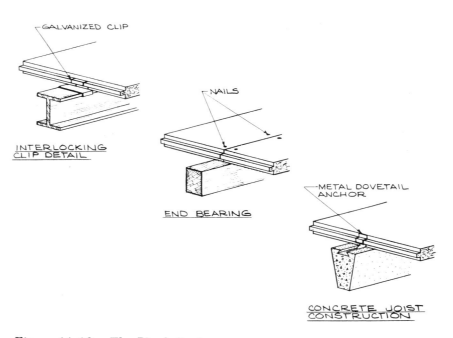

Figure 11-18. The Plank System

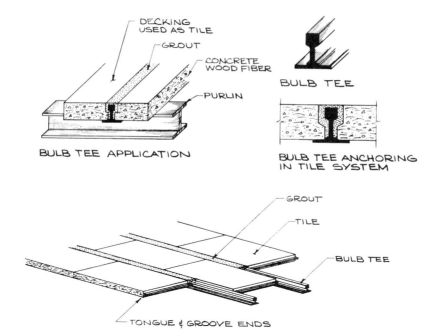

Figure 11-19.   The Bulb-Tee System

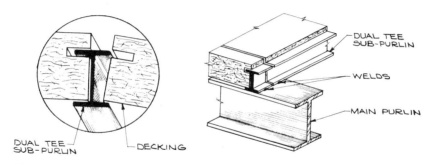

Figure 11-20.   Dual Tee Fastening Details

12

Brick

INTRODUCTION

This unit covers the manufacturing processes of clay and shale brick, the standard and special shapes generally used, physical characteristics, brick wall patterns, bonding and joints, and types of brick wall construction.

TECHNICAL INFORMATION

Manufacturing Methods

Clay and shale brick are manufactured by the soft-mud process, the stiff-mud process, and the dry-press process. In all these processes the bricks are formed to the desired shape, dried, and then burned in kilns. The chief difference in the finished brick is the method of molding.

The soft-mud process requires that the clays be mixed with sufficient water (20 to 30% moisture) so that they can be easily molded (see Figure 12-1). In order to stop the wet clay from sticking to the mold, either water or sand is used. When water is used, the finished brick is known as water-struck brick, and when sand is used, the brick is known as sand-struck brick.

The stiff-mud process requires that the clays be mixed with only sufficient water (12 to 15% moisture) so that they can be extruded through a die (see Figure 12-2). The brick sizes are cut by using tightly stretched wires. If the ends are cut, the brick is known as end-cut and if the sides are cut, the brick is known as side-cut. Both types are called wire-cut brick.

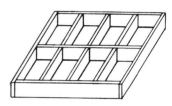

MOLD FOR BRICK

Figure 12-1. Soft Mud Process

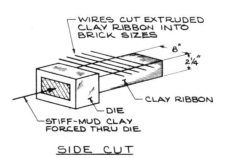

SIDE CUT   END CUT

Figure 12-2. Brick Produced by the Extrusion Method

In the dry-press process the clays are mixed with only enough water (7 to 10% moisture) so that they are almost in a dry state. Then they are molded into shapes with high pressure (see Figure 12-3).

Brick that is first pressed into oversize molds and then re-pressed in molds of the current size of brick is called re-pressed brick.

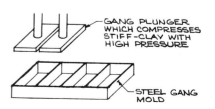

Figure 12-3. Dry Press Process

By varying the types of clays and introducing various metal oxides, various colors are produced.

Table 12-1. Size of Common, Face, and Modular Brick

Type of Brick	Depth	Width	Length
Common and face brick*	$2\frac{1}{4}''$ not including mortar joint	$3\frac{3}{4}''$ not including mortar joint	$8''$ not including mortar joint
Modular common and face brick	$2''$, $2\frac{2}{3}''$, $4''$, and $5\frac{1}{3}''$ including mortar joint	$4''$ including mortar joint	$8''$ and $12''$ including mortar joint

*Permissible variables are: plus or minus $\frac{1}{16}''$ in depth, $\frac{1}{8}''$ in width, and $\frac{1}{4}''$ in length

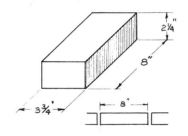 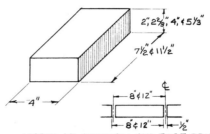

COMMON AND FACE BRICK—DIMENSIONS DO NOT INCLUDE MORTAR JOINTS

MODULAR COMMON AND FACE BRICK—DIMENSIONS INCLUDE ONE MORTAR JOINT OR TWO HALF JOINTS

Figure 12-4. Brick

Table 12-2. Physical Requirements of Common and Face Brick

Grade	Minimum Compressive Strength (gross area)	Maximum Water Absorption by 5-hr. Boiling (percent)	Maximum Saturation Coefficient*
SW	2500 PSI	20%	0.80
MW	2200 PSI	25%	0.90
NW	1250 PSI	no limit	no limit

*The saturation coefficient is the ratio of absorption by 24-hr. submersion in cold water to that after 5-hr. submersion in boiling water.

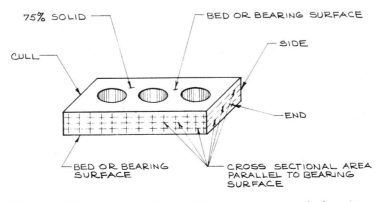

Figure 12-5. Hollow Brick (showing names of the six surfaces of brick)

Types and Grades of Brick

The types of brick generally used in building construction are common and face brick. Common brick is available in three grades:

Grade SW: used where heavy rain and freezing are encountered.
Grade MW: used where rain and freezing are not extreme.
Grade NW: used where there is minimum rain and freezing.

Face brick is available in SW and MW grades similar to common brick and three other grades based upon factors affecting the appearance:

Grade FBX: high degree of mechanical perfection and narrow color range.
Grade FBS: variations in mechanical perfection and wide color range.
Grade FBA: nonuniformity in size, color, and texture of brick units.

Size of Brick

Most common and face brick are manufactured in standard sizes with minor variations in dimension. The United States Bureau of Standards recommends that brick sizes should be as shown in Table 12-1 and Figure 12-4. All bricks vary slightly in their dimensions because of the shrinkage of the clays in burning.

Physical Requirements

The various grades of common and face brick (including FBX, FBS, and FBA) are required to meet exact physical requirements (see Table 12-2).

Most brick today is made with round or rectangular holes because these voids allow the brick to be more evenly burned throughout in the kiln.

In order to meet the physical requirements for the various grades of brick as established by the American Society for Testing Materials (ASTM), the definition of a brick is a unit whose net cross-sectional area in every plane parallel to the bearing surface is 75% or more of its gross cross-sectional area measured in the same plane. A solid brick measuring 3-3/4 × 7-3/4 in. = 29.06 sq. in. of bearing surface. With holes or voids, the cross-sectional area of the brick should be 75% of the total area, or 21.79 sq. in. The difference of 7.27 sq. in. is that of the holes or voids (see Figure 12-5).

Brick may be named or identified by a particular placement within a wall. Ordinarily, as one views a brick wall, the horizontal brick are called *stretchers*. When the brick-lengths are placed at right angles to the length of the wall, the brick are called *headers*. Similarly, others placed standing on end are known as *soldiers*. Figure 12-6 illustrates the various methods of placing brick.

Most brick manufacturers make a limited number of special-shaped brick such as bullnose, bullnose stretcher and header, bullnose starter (left or right), and internal bullnose return (see Figure 12-7).

Brick Bonds

Brick bonds are used primarily to assist in the strengthening of the brick wall. By bonding brick, various patterns or designs are created, of which some of the more common and better known are shown in Figure 12-8. These are the *common header bond*, where a header course is placed every 5th, 6th, or 7th course, the *common stretcher-header bond*, the *all-stretcher bond*, which is reinforced every sixth course with metal ties placed between the joints, and the *stacked bond*, with similar metal reinforcings every sixth course.

Brick hardness depends on the degree of burning received in the kiln. After burning, brick is sorted into the three following categories:

1. Brick nearest the fire — these are overburned, usually warped and discolored, and used where a decorative type

brick is desired. They are known as "*clinker brick.*"

2. Brick with the right amount of burning — known as "*hard-burned brick*" and used for general brick work.

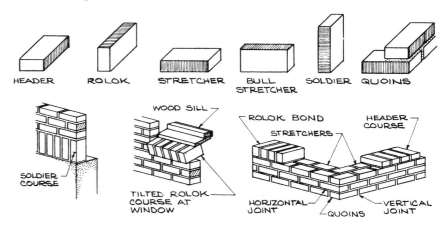

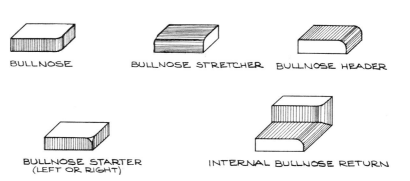

Figure 12-6. Placement of Brick

Figure 12-7. Special Shapes

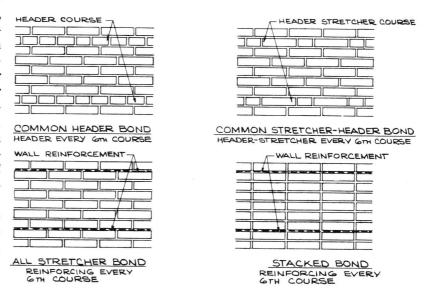

Figure 12-8. Types of Brick Bonds Generally Used Today in Building Construction

3. Under-burned brick — known as *"soft brick"* and used for back-up or in areas where strength is not of prime importance.

In building construction today most brick is used as a veneer with a masonry type back-up, usually concrete block, and therefore many brick sizes are dimensioned so that the coursing of brick coordinates with the coursing of the back-up material (see Figure 12-9).

There are other types of brick used to create decorative walls where the horizontal coursing is the dominant design factor. Two of such types are the *Roman*, 1-1/2 × 11-3/4 in., and the *Norman*, 3-3/4 × 11-5/8 in. (See Figure 12-10.)

The standard brick, 2-1/4 × 7-3/4 × 3-3/4 in., with 3/8 or 1/2-in. mortar joints, is generally used in the four common types of wall constructions shown in Figure 12-11. These are the plain 4-in. brick veneer wall; the 10-in. cavity wall; the 8-in. solid wall; and the 12-in. solid wall.

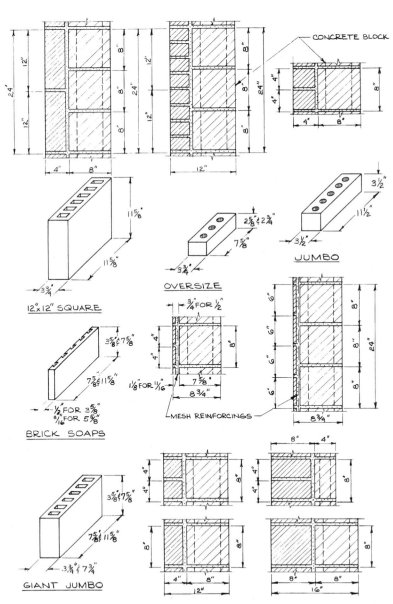

Figure 12-9.   Brick Veneer

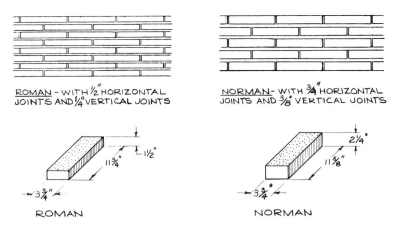

ROMAN - WITH ½" HORIZONTAL JOINTS AND ¼" VERTICAL JOINTS

NORMAN - WITH ¾" HORIZONTAL JOINTS AND ⅜" VERTICAL JOINTS

ROMAN

NORMAN

Figure 12-10. Decorative Brick

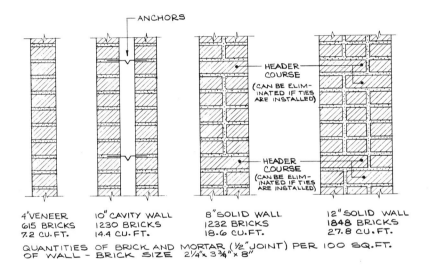

4" VENEER
615 BRICKS
7.2 CU. FT.

10" CAVITY WALL
1230 BRICKS
14.4 CU. FT.

8" SOLID WALL
1232 BRICKS
18.6 CU. FT.

12" SOLID WALL
1848 BRICKS
27.8 CU. FT.

QUANTITIES OF BRICK AND MORTAR (½" JOINT) PER 100 SQ. FT. OF WALL - BRICK SIZE 2¼" x 3¾" x 8"

Figure 12-11. Types of Brick Wall Construction

REVIEW EXAMINATION

1. Name the three processes used in manufacturing brick.

2. If the molds are made wet in order to stop the wet clay from sticking, what are the bricks called?

3. If the molds are sanded, what are the bricks called?

4. Give the dimensions of common and face brick.

5. At what courses may header bonds be located?

6. The hardness of brick depends upon what?

7. What are the three grades of brick that meet exact physical requirements?

8. The compressive strength in PSI of an MW brick is greater than that of an SW brick — true or false?

9. Name the six sides of a brick.

10. In reference to hollow brick, what must the net cross-sectional area in every plane parallel to the bearing surface be?

ASSIGNMENT

1. Make a freehand sketch to scale showing four brick wall types.

2. Make a freehand sketch to scale showing six various placements of brick.

3. Make a freehand sketch showing a hollow brick and the names of six sides of brick.

SUPPLEMENTARY INFORMATION

Water Absorption of Brick

The rate of absorption plays an important role in the bonding of the brick to the mortar in the joint. If the brick absorbs water from the mortar too quickly a poor bond will result, causing leaks and other damage. Tests have indicated that brick with absorption rates in excess of 0.7 oz. per min. must be wetted sufficiently to prevent the rapid absorption of the water from the mortar and thereby allowing the time required for the proper setting of the mortar.

All brick should be laid in temperatures above 40° F., and in very high temperatures such as 90° and over, caution should be taken that the brick is sufficiently wetted so that the mortar does not set too fast.

In all building construction brick work, the horizontal and vertical brick coursing should be checked for window openings, door openings, floor-to-floor heights, etc.

Other Types of Brick

Glazed facing brick. Brick that receive a glaze on one face are known as "glazed facing brick." The glaze is applied directly after the brick is removed from the kiln, and is again burned until the glaze is fused to the brick. Because different color glazes may be applied, a large variety of colored facing brick can be made. Glazes may have a mat or gloss finish.

Cement brick. This type is manufactured in the same sizes as common and face brick and there are two grades, B and C (see Table 12-3). Cement brick is manufactured from a controlled mixture of portland cement and aggregates (sand, crushed rock, gravel, blast-furnace slag, or burned clay or shale). Grade B is frost-resistant and Grade C is used for back-up and interior work. Various colors and textures are available.

Sand-lime brick. Manufactured by the semi-dry method in presses and

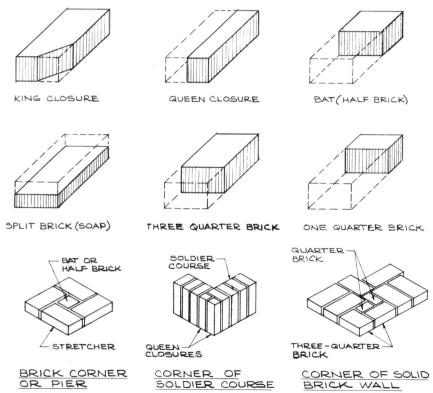

Figure 12-12. Shapes, Names, and Placement of Brick

cured in autoclaves for 24 hrs. with steam, this brick is made from 5 to 10% hydrated lime and sand.

Sand-lime brick is graded the same as common and face brick and has good frost resistance and better fire resistance than common or face brick. Its natural color is a pearl-gray, but other colors are available.

Adobe brick. Made from natural clays to which straw is added to prevent shrinkage and cracks. Both are mixed to a consistency for molding and then placed in the sun for two weeks to dry. Adobe brick should always be protected from moisture by large roof overhangs and concrete foundations to prevent capillary action.

Shapes and names of cut brick. Figure 12-12 illustrates brick that are cut or split into other shapes and are used to fulfill a definite need in brick construction.

Table 12-3. Cement Brick Strength Requirements

Grade	Compressive Strength (PSI)		Modulus of Rupture (PSI)	
	5 Tests Mean	Minimum Individual	5 Tests Mean	Minimum Individual
B	2250	1500	450	300
C	1250	1000	300	200

13

Building with Brick

INTRODUCTION

This unit provides information on brick mortars and joints, strength of brick walls, isolated piers and piers within a wall, expansion joints and expansion joint materials, brick coursing, accessories for brickwork, and illustrations of wall sections showing various methods of building with brick.

TECHNICAL INFORMATION

Mortars

The mortars used for brickwork are divided into three categories:

(1) *Type A* — rigorous exposure, load-bearing, and below grade.

(2) *Type B* — above grade, severe exposures, and interior load-bearing.

(3) *Type C* — nonbearing, bearing where compressive strength does not exceed 100 PSI.

Mortars are available ready-mixed (in bags), premixed (mixed at plant and transported to site), and job-mixed (ingredients are mixed by hand or in a cement mixer at site). Mortars for large construction are controlled by laboratory testing to meet the requirements known as *property specifications* as shown in Table 13-1, or by predetermined proportions by volume, known as *proportion specification*, as shown in Table 13-2.

Sand for mortar is generally white and is graded by size of joint (see Table 13-3).

Table 13-1. Property Specifications for Mortar Types A, B, and C

Type of Mortar	Compressive Strength in PSI (average of three 2" cubes)	
	7 days	28 days
A	1500	2500
B	550	900
C	200	350

Table 13-2. Mortar Proportion Specification by Volume and Weight

Type of Mortar	Cement Portland 94 lbs. = 1 cu.ft.	Lime Hydrated 40 lbs. = 1 cu. ft.	Lime putty 40 lbs. = 1 cu. ft.	Aggregate Sand 80 lbs. = 1 cu. ft.
A	1 cu. ft. or 94 lbs.	1/4 cu. ft. or 10 lbs.	1/4 cu. ft. or 10 lbs.	3 cu. ft. or 240 lbs.
B	1 cu. ft. or 94 lbs.	1 cu. ft. or 40 lbs.	1 cu. ft. or 40 lbs.	6 cu. ft. or 480 lbs.
C	1 cu. ft. or 94 lbs.	1 cu. ft. or 40 lbs.	2 cu. ft. or 80 lbs.	9 cu. ft. or 720 lbs.

Table 13-3. Sand Graded for Mortar Joints Larger than 1/4"

Screen No.	Size of Opening (in inches)	Percentage of Sand Retained (by weight)	
		Maximum	Minimum
8	0.097	10	0
16	0.049	40	15
30	0.0232	65	35
50	0.0117	85	75
100	0.0059	98	95

For small jobs there exist pre-mixed mortars that can be obtained usually from any local building supply store. For the small portable mixer the following mortar is used: 1–2–4; namely, one part portland cement to 2 parts lime putty to 4 parts fine sand.

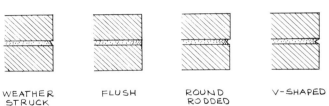

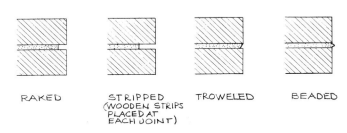

Figure 13-1. Mortar Joints

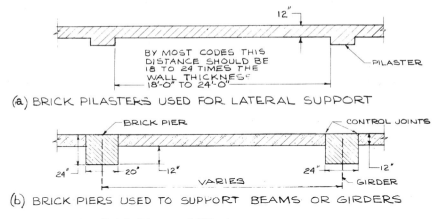

Figure 13-2. Brick Piers and Pilasters

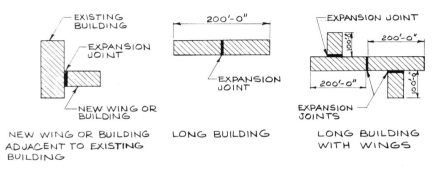

Figure 13-3. Expansion Joints

Mortar Joints

Mortar joints for brickwork are generally 3/8 and 1/2 in. Exterior joints should be of such a type that will easily shed water, whereas on the interior or in climates with little rainfall the joints do not have to shed water. The difference is illustrated in Figure 13-1.

Strength of Brick Walls

The strength of brick walls is often influenced by workmanship. Walls categorized as "inspected," type A workmanship, using SW brick and type A mortar, will have a strength of one-half the compressive strength of the brick. Walls categorized as "ordinary," type B workmanship, have a strength of two-thirds that of type A workmanship according to standard tests.

Piers

Brick piers are used to laterally support walls (see Figure 13-2, A). Where brick piers are used to support girders or roof trusses, their dimensions should be sufficient so that the allowable compressive working stress of the brick masonry is not exceeded (see Figure 13-2, B). The formula used is:

$$A = P/fc$$

where

A = required area in sq. in.
P = load
fc = allowable compressive working stress in pounds per sq. in.

Example

A roof load on a roof truss is 480,000 lbs., or 480 kips (a kip is 1000 lbs.). This load is supported by brick masonry which when inspected was classed as type A workmanship. The compressive strength of brick is 2500 PSI. Because the allowable working stress is taken as one-half of the compressive strength of the material, the allowable working stress is therefore 1250 PSI.

Therefore,

$$A = \frac{P}{fc}$$

$$A = \frac{480,000}{1250} = 384 \text{ sq. in.}$$

This would provide for a pier dimension of 16 × 24 in. For safety factor reasons, a somewhat larger pier can be selected, such as 20 × 24 in., or 16 × 28 in. (see Figure 13-2, B).

Isolated brick piers are calculated by using the same formula, but their heights are limited by codes to six to ten times their least dimensions.

Expansion Joints

All large buildings over 200'-0" long or buildings with one or more wings require at least one expansion joint whether they are of masonry, concrete, or steel frame construction. When new additions are added to existing buildings, they must be separated by expansion joints (see Figure 13-3).

Example

A brick masonry building is 400'-0" long. The coefficient of linear expansion of brick is 0.0000031 in. per degree Fahrenheit. Using the formula:

Length in in. × temperature range × coefficient of expansion

or:

expansion = 400 × 12 × 100 × 0.0000031

where:

400 = length in ft. × 12 in. (conversion of ft. to in.)

100 = average temperature range (e.g., 0° to 100°) winter and summer extremes

Total expansion of building is 1.488 in. or 1-1/2 in. On the basis of this information, an expansion joint of 1 in. would be necessary at the center of the building as shown in Figure 13-4. These expansion

joints completely isolate one part of the building from the other, including foundations, exterior walls, partitions, floor slabs, ceilings, parapets and roof construction, and roofing (see Figure 13-5).

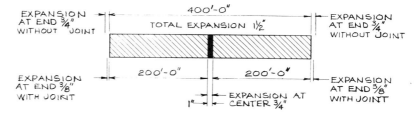

Figure 13-4. Expansion of a Building

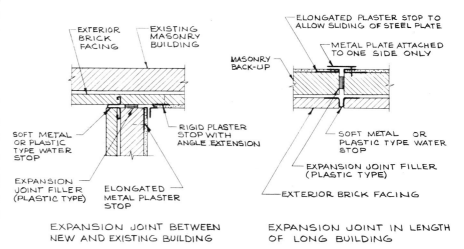

Figure 13-5. Expansion Joints

Table 13-4. Vertical Coursing of Standard and Modular Brick

$2\frac{1}{4}" + \frac{3}{8}"$ joint		$2\frac{1}{4}" + \frac{1}{2}"$ joint		$2\frac{2}{3}"$ Modular (includes joint)		4" Modular (includes joint)	
Course	Height	Course	Height	Course	Height	Course	Height
1	$2\frac{5}{8}"$	1	$2\frac{3}{4}"$	1	$2\frac{2}{3}"$	1	4"
2	$5\frac{1}{4}"$	2	$5\frac{1}{2}"$	2	$5\frac{1}{3}"$	2	8"
3	$7\frac{7}{8}"$	3	$8\frac{1}{4}"$	3	8"	3	12"
4	$10\frac{1}{2}"$	4	11"	4	$10\frac{2}{3}"$	4	1'-4"
5	$1'-1\frac{1}{8}"$	5	$1'-1\frac{3}{4}"$	5	$1'-1\frac{1}{3}"$	5	1'-8"
6	$1'-3\frac{3}{4}"$	6	$1'-4\frac{1}{2}"$	6	1'-4"	6	2'-0"
7	$1'-6\frac{3}{8}"$	7	$1'-7\frac{1}{4}"$	7	$1'-6\frac{2}{3}"$	7	2'-4"
8	1'-9"	8	1'-10"	8	$1'-9\frac{1}{3}"$	8	2'-8"
9	$1'-11\frac{5}{8}"$	9	$2'-0\frac{3}{4}"$	9	2'-0"	9	3'-0"
10	$2'-2\frac{1}{4}"$	10	$2'-3\frac{1}{2}"$				

Fillers and waterstops for expansion joints. Fillers are manufactured in three types: premolded, mastic, and metal.

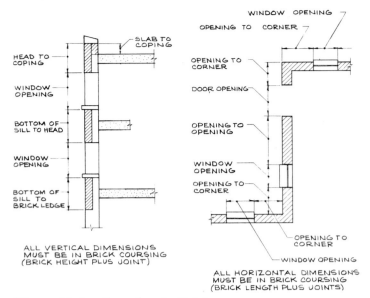

ALL VERTICAL DIMENSIONS
MUST BE IN BRICK COURSING
(BRICK HEIGHT PLUS JOINT)

ALL HORIZONTAL DIMENSIONS
MUST BE IN BRICK COURSING
(BRICK LENGTH PLUS JOINTS)

Figure 13-6.   Openings in Brick

Pre-molded type:
1. Vinyl, neoprene, and other resilient-type plastics
2. Cork
3. Rubber and synthetic rubbers
4. Asphalt-impregnated materials

Mastic type:
1. Epoxy
2. Asphalt
3. Synthetic rubbers
4. Silicones

Metal type: Bellows (soft copper or stainless steel)

Coursing

In brick construction work, both vertical and horizontal coursing of the brick is of utmost importance. In vertical coursing, window and door openings must be carefully calculated by brick coursing because most windows and doors are manufactured in stock sizes, both in height and width, as shown in Figure 13-6 and Table 13-4.

Examples of Wall Sections

Brick veneer with frame backing. This section (see Figure 13-7) illustrates how the brick veneer is started on a concrete ledge somewhat below grade. The veneer wall is bonded to the sheathing by corrugated metal ties spaced 16 in. OC horizontally and vertically, and is carried up to or beyond the soffit line, or as high as the overhanging rafters will permit. The brick veneer in this construction is non-load-bearing and is therefore used only for appearance.

Brick cavity wall with steel frame and open web steel joists. This wall section (see Figure 13-8) illustrates how the brick veneer wall is anchored to concrete block by means of cavity wall ties and by dovetail cavity wall anchors. The spandrel steel sections support open-web steel floor and roof joists,

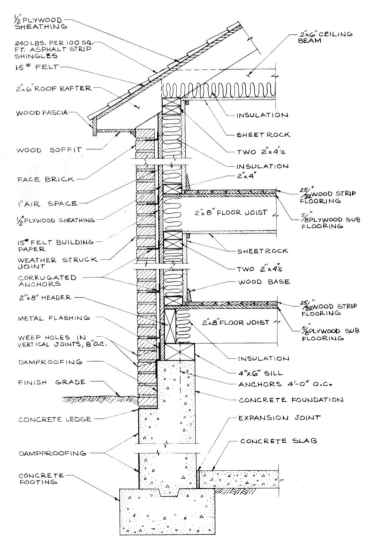

Figure 13-7.   Brick Veneer on Wood Frame

which are covered by corrugated steel sheets acting as permanent forms for the 2-1/2-in. reinforced concrete slab. Note the spandrel flashing and the weep-holes for the elimination of condensation from the cavity.

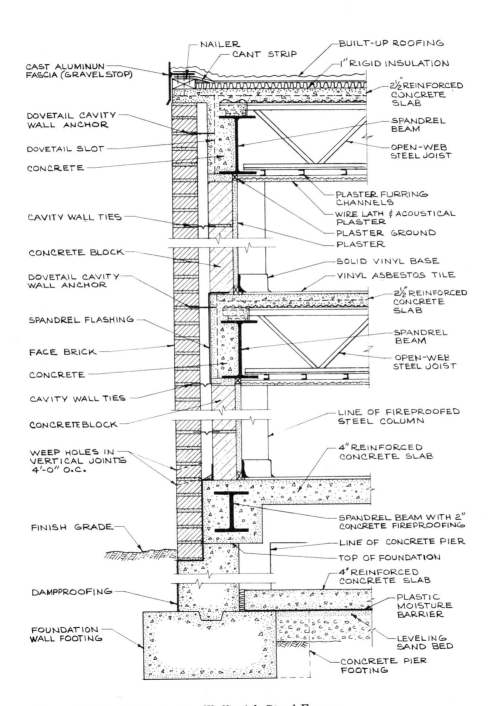

Figure 13-8. Brick Cavity Wall with Steel Frame and Open-Web Steel Joists

1. Name the three types of mortars used in brickwork.

2. For large construction, mortars are controlled by what two methods?

3. What are the generally used sizes of mortar joints?

4. In the formula A = P/fc, what is A? P? fc?

5. When you add an addition or a new wing to a building, what do you install at their connection?

6. Name the three types of fillers and waterstops that are used in expansion joints.

7. In a building using face brick on the exterior, what is of utmost importance in the way of calculations that must be made?

8. Name two types of brick joints that shed water.

9. Name two types of brick joints that do not shed water.

10. Over what length should a building have an expansion joint?

ASSIGNMENT

1. Draw a freehand sketch showing where expansion joints are required in buildings.

2. Draw a freehand sketch to scale showing eight types of brick joints.

SUPPLEMENTARY INFORMATION

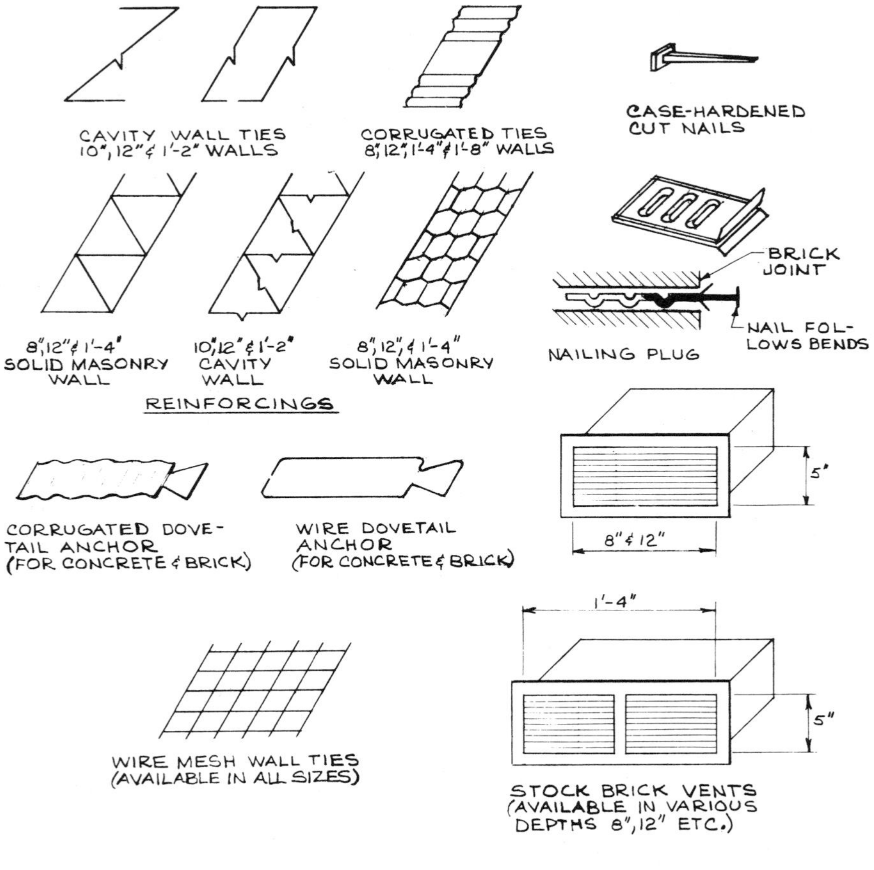

Figure 13-9. Brick Work Accessories

Efflorescence

This is a white stain that appears on brickwork and is caused by soluble salts which are in the material and brought to the surface of the brick by moisture. In order to overcome efflorescence, it is necessary to check the type of brick (most brick now produced do not have soluble salts), the type of mortar and, most important, the correct installation of the flashing.

Accessories

Accessories for brickwork are of utmost importance in building construction. They consist of cavity wall ties, reinforcing, nailing plugs, anchors and ties, nails, and various control and expansion joint materials. These are illustrated in Figures 13-9 and 13-10.

Special Mortars and Mortar Admixtures

Plastic mortar. There is an epoxy cement mortar which is now being used for brickwork. This mortar is applied in a ribbon with a type of caulking gun — it sets in 24 hours and reaches full strength in 72 hours. It not only is equal to conventional mortars in compressive strength but also has the advantage of great tensile strength, whereas conventional mortar has very small tensile strength. With this type of mortar, prefabricated brick and concrete panels 12′–0″ long by 8′–0″ high can be made and then lifted into place in the building.

Mortar admixtures. These consist of accelerators, plasticizing agents, waterproofing agents, and color pigments. Admixtures come in powders, paste, and liquid form and are usually patented and sold under trade names.

Grout. This refers to mortars containing finely ground iron particles and hardening elements mixed with cement. Grout sets rapidly, does not shrink or expand, and forms a hard, strong, durable material. It is used for setting steel, columns, etc., onto concrete or masonry walls, and also for heavy machinery installations.

Fire Brick

This type is made of fire clay that contains 50% alumina and is used for fireplaces where intense heat is present. Fire brick is installed with a fire clay-type of mortar. There are other kinds of special brick such as refractory brick (used where very high temperatures are present), flooring and paving brick (see Figure 13-11), acid brick (used where acids are present), and brick for reinforced brick walls (see Figure 13-12).

More Examples of Wall Sections

Brick with concrete block back-up. This is shown in Figure 13-13. The brick veneer and the concrete block backing in this wall section are tied by corrugated flat wall ties. The ends of the wood floor joists are fire-cut and bear a minimum of 4 in. on filled-in solid blocks. The wall is provided on its inside surface

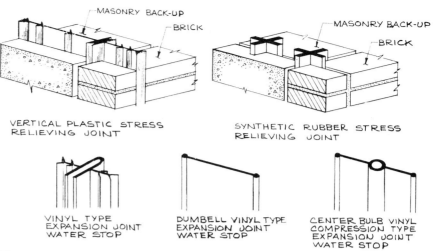

Figure 13-10. *Control and Expansion Joint Materials*

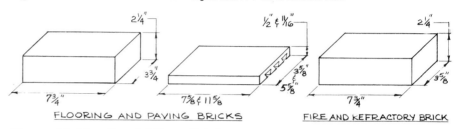

Figure 13-11. *Special Type Bricks*

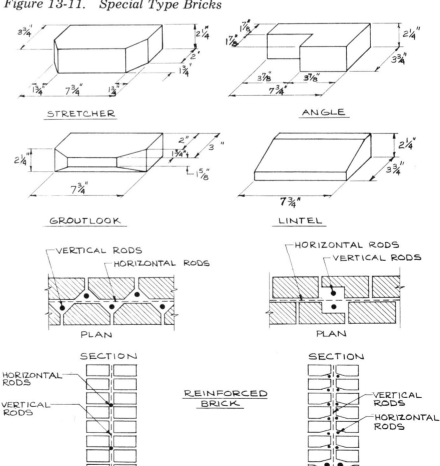

Figure 13-12. *Special Brick*

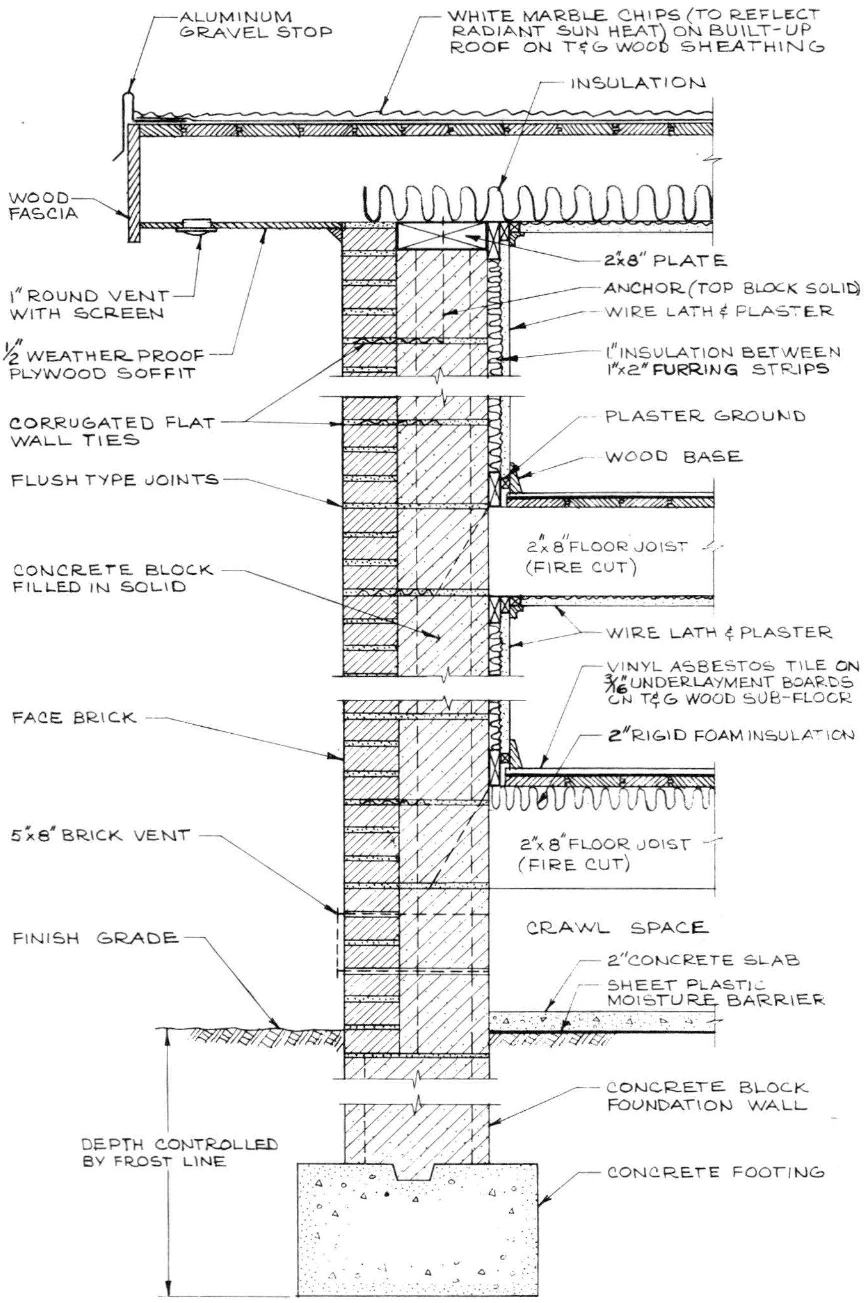
ALUMINUM GRAVEL STOP
WHITE MARBLE CHIPS (TO REFLECT RADIANT SUN HEAT) ON BUILT-UP ROOF ON T&G WOOD SHEATHING
INSULATION
WOOD FASCIA
1" ROUND VENT WITH SCREEN
½" WEATHER PROOF PLYWOOD SOFFIT
CORRUGATED FLAT WALL TIES
FLUSH TYPE JOINTS
CONCRETE BLOCK FILLED IN SOLID
FACE BRICK
5"x8" BRICK VENT
FINISH GRADE
DEPTH CONTROLLED BY FROST LINE
2"x8" PLATE
ANCHOR (TOP BLOCK SOLID)
WIRE LATH & PLASTER
1" INSULATION BETWEEN 1"x2" FURRING STRIPS
PLASTER GROUND
WOOD BASE
2"x8" FLOOR JOIST (FIRE CUT)
WIRE LATH & PLASTER
VINYL ASBESTOS TILE ON 3/16" UNDERLAYMENT BOARDS ON T&G WOOD SUB-FLOOR
2" RIGID FOAM INSULATION
2"x8" FLOOR JOIST (FIRE CUT)
CRAWL SPACE
2" CONCRETE SLAB
SHEET PLASTIC MOISTURE BARRIER
CONCRETE BLOCK FOUNDATION WALL
CONCRETE FOOTING

Figure 13-13. Brick with Concrete Block Back-Up

with 1 × 2-in. furring strips and 1-in. insulation. The finish is wire lath and plaster. The relatively flat roof is covered with white marble chips (to reflect radiant sun heat) on a built-up roof on tongue-and-groove wood roofing boards.

A brick wall with reinforced concrete. This is illustrated in Figure 13-14. The brick wall with concrete block backing is integrated with the concrete floor and roof slabs. The roof slab, for example, is cast to form part of the wall lintel and cap over a window opening. The roof is insulated with 1-1/2-in. rigid insulation and a pitched fill provides for proper roof drainage. Note the corrugated wall anchors, the spandrel flashing, the base and cap flashing, and the expansion joints where the concrete sidewalk and the cellar slab meets the wall.

Mortars for Small Construction

For small construction work, the materials used for mortars are usually in the following proportions:

1.0 part cement
3.0 parts sand
0.1 part of hydrated lime

Ready-mixed mortars for small work, furnished in paper bags, are available from building materials suppliers.

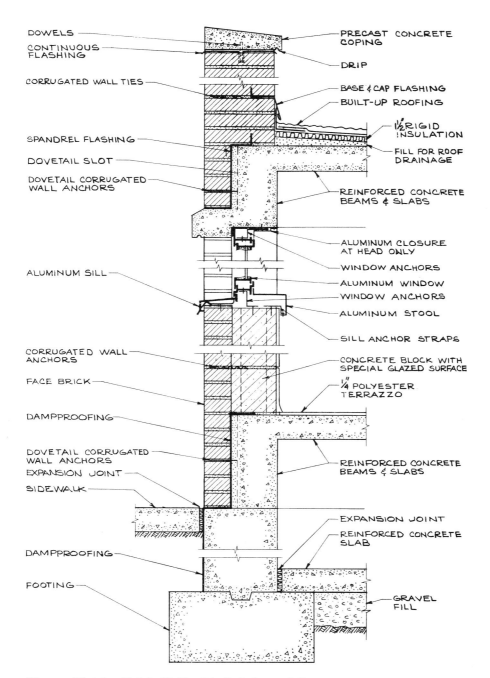

DOWELS

CONTINUOUS FLASHING

CORRUGATED WALL TIES

SPANDREL FLASHING

DOVETAIL SLOT

DOVETAIL CORRUGATED WALL ANCHORS

ALUMINUM SILL

CORRUGATED WALL ANCHORS

FACE BRICK

DAMPPROOFING

DOVETAIL CORRUGATED WALL ANCHORS

EXPANSION JOINT

SIDEWALK

DAMPPROOFING

FOOTING

PRECAST CONCRETE COPING

DRIP

BASE & CAP FLASHING

BUILT-UP ROOFING

1½" RIGID INSULATION

FILL FOR ROOF DRAINAGE

REINFORCED CONCRETE BEAMS & SLABS

ALUMINUM CLOSURE AT HEAD ONLY

WINDOW ANCHORS

ALUMINUM WINDOW

WINDOW ANCHORS

ALUMINUM STOOL

SILL ANCHOR STRAPS

CONCRETE BLOCK WITH SPECIAL GLAZED SURFACE

¼" POLYESTER TERRAZZO

REINFORCED CONCRETE BEAMS & SLABS

EXPANSION JOINT

REINFORCED CONCRETE SLAB

GRAVEL FILL

Figure 13-14. Brick Wall with Reinforced Concrete

14

Other Masonry Units

INTRODUCTION

This unit covers the various masonry-type units other than brick, such as structural load-bearing and non-load-bearing clay wall tile, facing tile, concrete block, concrete block with special surface finish, terra cotta, and gypsum block. These materials are discussed with regard to their sizes and uses, including information on their load-bearing capacities, fire ratings, and their general material composition.

TECHNICAL INFORMATION

By definition, a masonry unit is a rectangular-shaped block whose net cross-sectional area in any plane parallel to the bearing surface is less than 75% of its gross cross-sectional area measured in the same plane. For example: The masonry unit shown in Figure 14-1 measures 3-3/4 × 7-3/4 in. Therefore, its area is:

3.75 × 7.75 = 29.06 sq. in.

The area of two holes is:

2.25 × 2.75 × 2 = <u>12.38</u> sq. in.
Net bearing cross-sectional area
= 16.68 sq. in.
75% of 29.06 = 21.8 sq. in.

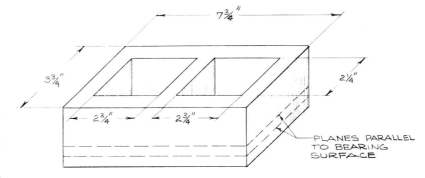

Figure 14-1. Definition of a Masonry Unit

Because the net bearing area of the block is less than 75% of total area, the block by definition is a masonry unit.

Structural Clay Tile

This type of tile is a hollow masonry unit used in building construction for load-bearing and non-load-bearing walls, fireproofing, back-up, and furring, but not for exterior or interior surface finish (see Figure 14-2).

Load-bearing structural tile is manufactured in two grades: LBX for general masonry work, and LB for masonry where there is a 3-in. or more exterior facing. Non-load-bearing, fireproofing, and furring

115

tile are manufactured in one grade only. In the building construction

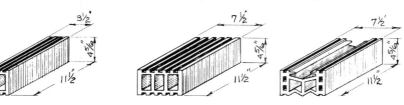

STRUCTURAL CLAY WALL TILE — LOAD BEARING

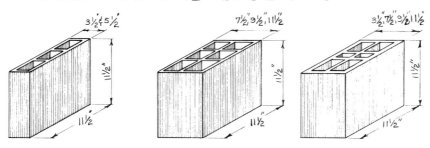

STRUCTURAL CLAY WALL TILE-LOAD BEARING-END CONSTRUCTION

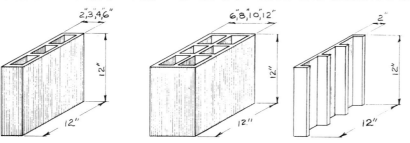

STRUCTURAL CLAY WALL TILE — NON LOAD BEARING
PARTITION AND FURRING TILE

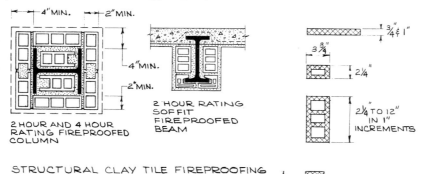

2 HOUR AND 4 HOUR RATING FIREPROOFED COLUMN

2 HOUR RATING SOFFIT FIREPROOFED BEAM

STRUCTURAL CLAY TILE FIREPROOFING

Figure 14-2. Structural Clay Tile

Table 14-1. Compressive Strength of Load-bearing Structural Facing Tile

Class	Minimum Compressive Strength of Single Unit (PSI gross area)	
	End Construction	Side Construction
Standard	1000	500
Special duty	2000	1000

field, these structural tiles are still used extensively but are not as important as they formerly were because of new methods of fire-proofing steel, the widespread use of prefabricated partitions, and the development of various types of concrete block and new methods of finishing surfaces.

Load-bearing Structural Facing Tile

This type of tile is very extensively used in building construction for interior partitions, for back-up for exterior walls and, in combination with other masonry units, for partitions. Load-bearing structural facing tiles are manufactured in two classes, *standard* and *special duty* (see Table 14-1).

They are available in a wide range of colors, either bright or matte finish, and in various textures including an acoustical type. The sizes have been standardized. These tiles are manufactured for both end and side construction. Also, special shapes for cove bases, sills, caps, lintels, miters, internal and external corners, jambs, and starters are standardized and readily available (see Figure 14-3). This type of tile has a most important additional feature in that it is available with finish surfaces on both sides.

Load-bearing structural facing tile is generally called *structural facing tile*, or *SFT*, and is graded in two categories: FTX, with a high degree of mechanical perfection and narrow range in color variation — these are used where low water absorption, easy maintenance, and resistance to stain are required; and FTS, with good mechanical perfection and wide range in color variation.

Concrete Block

Concrete block is perhaps the most widely used masonry unit material in building construction today, other than brick. Because of the ease in manufacture and its light weight, strength, and fire-proofing qualities, it has supplanted many other types of masonry units. It is used for foundations, back-up, fireproofing, load-bearing and non-load-bearing walls and partitions and, with special surface finishes, for exterior and interior areas where easy maintenance and stain resistance are required (see Figure 14-4).

Concrete block is manufactured not only in 7-3/4-in. heights but also in 3-1/2-in. and 5-in. heights in order to meet the heights of brick and structural facing tile. Load-bearing concrete block is manufactured either solid or cored, and all non-load-bearing block is cored. Load-bearing concrete block is graded in two categories, *grade A* and *grade B* (see Table 14-2).

Concrete Block with Surface Finish (Glazed Masonry Units)

Concrete block with colored, durable, and easily maintained surface is used for back-up, load-bearing and non-load-bearing walls and partitions, and where a hard, durable, easily maintained finish surface is required (see Figure 14-5). It is available with both faces of the block having a colored, durable finish.

Special shapes are available such as caps, lintel blocks, bull noses, header blocks, and thin caps and bases.

This type of concrete block is manufactured so that the durable surface is slightly larger than the block to which it is applied so that all joints are 1/4 in. (see Figure 14-6).

Terra Cotta (Ceramic Veneer)

This type of masonry unit was very widely used in building construction in the past. Recently has it again become popular, its name having been changed to ceramic veneer. This material has the advan-

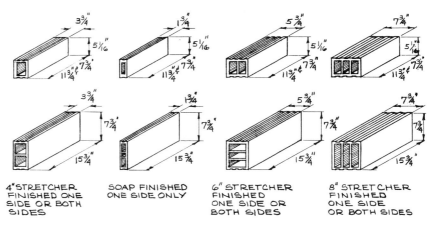

STRUCTURAL CLAY FACING TILE—LOAD BEARING—SIDE CONSTRUCTION (DIFFERENT MANUFACTURERS SUPPLY BOTH SIDE AND END CONSTRUCTION TYPES)

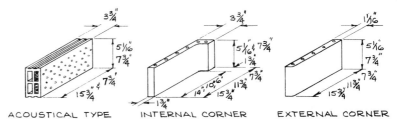

Figure 14-3. Structural Clay Facing Tile—Load Bearing

Table 14-2. Load-bearing Concrete Block

Type of Block	Thickness of Shell (minimum)	Grade	Compressive Strength of Average Gross Area (PSI)	
			5 blocks average	1 block
Hollow load-bearing	1 1/2″ or over	A	1000	800
	1 1/4″ or over	B	700	600
Hollow non-load-bearing	not less than 1/2″		350	300
Solid load-bearing		A	1800	1600
		B	1200	1000

LOAD-BEARING AND NON-LOAD-BEARING CONCRETE BLOCK

Figure 14-4. Concrete Block

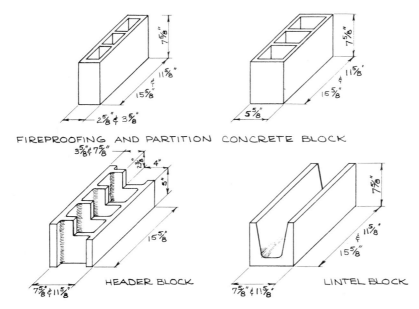

FIREPROOFING AND PARTITION CONCRETE BLOCK

HEADER BLOCK   LINTEL BLOCK

Figure 14-4.   *Concrete Block (continued)*

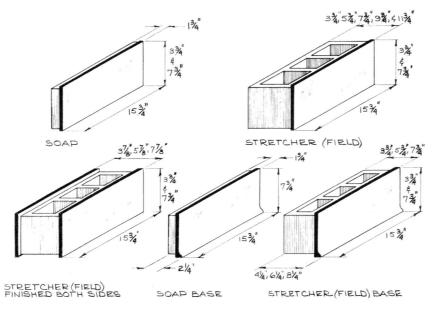

SOAP   STRETCHER (FIELD)

STRETCHER (FIELD)
FINISHED BOTH SIDES   SOAP BASE   STRETCHER (FIELD) BASE

Figure 14-5.   *Concrete Block with Surface Finish
(Glazed Masonry Unit)*

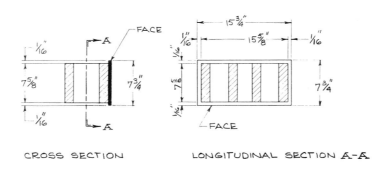

CROSS SECTION   LONGITUDINAL SECTION A-A

Figure 14-6.   *Concrete Block and Surface Finish*

---

tage of being a burned clay product and, more important, it is available in larger sizes, the maximum size being 2 ft.- 8-3/4 in. long × 4 ft.- 3 in. high and only 0′- 2-5/8 in. thick (see Figure 14-7). It is available in a limited range of colors but can be obtained in a wide range of colors with glazed finishes.

Gypsum Block

Gypsum blocks are manufactured from gypsum plaster (plaster of Paris) into solid and cored blocks which are used for fireproofing and non- load-bearing partitions where finish plaster will be applied. Gypsum, because of its chemical and physical make-up, has good fireproofing characteristics. For example, 2 in. of gypsum block has the same fire rating as 4 in. of concrete block (see Figure 14-8 and Table 14-3).

Table 14-3.   *Fire Ratings of Gypsum Block*

Type of Gypsum Block	With Plaster One or Both Sides	Fire Resistance (hours)
2″ solid for fireproofing steel	one side	4
3″ hollow for fireproofing steel	one side	4
3″ hollow partition	one side both sides	1 1/2 3
4″ hollow partition	one side both sides	3 4
6″ hollow partition	one side both sides	4 5

118

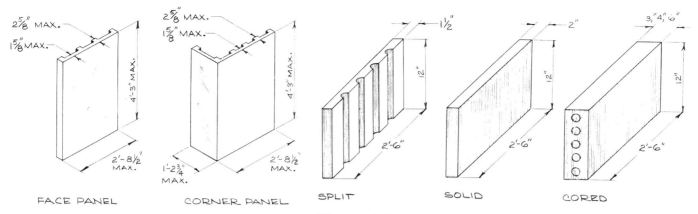

Figure 14-7. Ceramic Veneer (Terra Cotta)

Figure 14-8. Types of Gypsum Block

REVIEW EXAMINATION

1. A masonry unit's net cross-sectional area in any plane parallel to the bearing surface is less than what percentage of its gross cross-sectional area measured in the same plane.

2. Structural clay tile is used in building construction for what purposes?

3. Structural clay tile is installed in building construction by what two construction methods?

4. What are the two grades for load-bearing structural tile?

5. Load-bearing structural facing tile is graded in what two categories?

6. The load-bearing structural facing tile FTS grade has a high degree of mechanical perfection and a narrow range of color variation — true or false?

7. Structural facing tile cannot be manufactured with finish surfaces on both sides — true or false?

8. What are the advantages of using terra cotta (ceramic facings)?

9. Name four uses of concrete block in building construction.

10. In what two ways is load-bearing concrete block manufactured?

11. What are the two grades of load-bearing concrete block?

12. In what circumstances is concrete block with colored, durable, and easily maintained surface used in building construction?

13. What is gypsum block used for?

ASSIGNMENT

1. Draw a freehand sketch to scale showing a masonry unit.

2. Draw a freehand sketch to scale showing the various types of load-bearing structural facing tile (side construction).

SUPPLEMENTARY INFORMATION

Water absorption of concrete block. All concrete block is manufactured with a maximum water absorption of 1 lb. per cu. ft. of concrete, based on the average of 5 blocks per cu. ft. of concrete, each block weighing 15 lbs.

Special types of concrete block. With the development of concrete block with special surfaces, a new group of special concrete blocks is now manufactured, including concrete block with high acoustical value and a durable surface, block with a highly dense and polished surface, and block with a wide variety of surface textures incorporating marble chips, pebbles, and the like, and molded textures.

15

Building with Masonry Materials

INTRODUCTION

This unit covers building with other masonry materials — their limitations in height, strength, and load-bearing capacities, and methods of installation in building construction.

The masonry units other than brick include structural clay tile and facing tile, concrete block, special finished concrete block, terra cotta and gypsum block, and various types of mortar and joints used with these units.

TECHNICAL INFORMATION

Structural Clay Tile

Structural clay tile is manufactured in natural finish — that is, the color of the clay, which may be red, brownish, or yellowish. It is produced in two grades; namely, LBX and LB, each with different compressive strengths measured in PSI (see Table 15-1).

Load requirements. Structural clay tile is used for fireproofing such as that around steel columns, for wall furring, as a back-up in walls where a plaster or ceramic tile

Table 15-1. Load Requirements for Load-bearing Structural Clay Tile

Grade	Minimum Compressive Strength (PSI) for Single Unit	
	End Construction	Side Construction
LBX	1000	500
LB	700	500

is applied as a finish surface, but most of all as a load-bearing or non-load-bearing wall. Figure 15-1 illustrates structural clay tile for partition walls, one with a plaster finish on both sides and the other with a plaster finish on one side and a ceramic tile finish on the other. Height and length limitations of

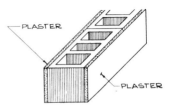
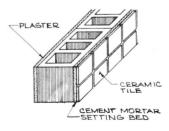

Figure 15-1. Structural Clay Tile Used for Load Bearing and Non-Load Bearing Partitions

Table 15-2. Fire Resistance Rating for Load-bearing Structural Clay Wall Tile for Wall Thicknesses as Shown

Type of Wall Construction	Combustible Material Framed into Wall (wood joists)				Noncombustible Materials Framed into Wall (steel joists)			
	4 hr.	3 hr.	2 hr.	1 hr.	4 hr.	3 hr.	2 hr.	1 hr.
Unplastered	16" 4 cells	16" 4 cells	12" 3 cells	12" 3 cells	12" 4 cells	12" 3 cells	12" 3 cells	8" 2 cells
Plastered (gypsum) one side	same as above	12" 3 cells	12" 2 cells	8" 2 cells	12" 3 cells	same as above	8" 2 cells	same as above
Plastered (gypsum) both sides)	same as above	same as above	same as above	same as above	same as above	8" 3 cells	same as above	same as above

121

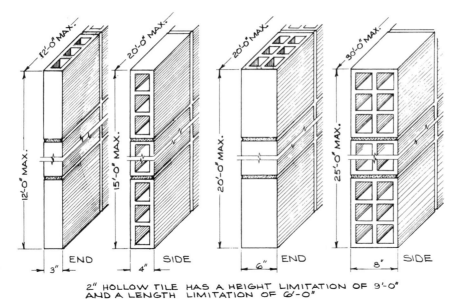

2" HOLLOW TILE HAS A HEIGHT LIMITATION OF 9'-0"
AND A LENGTH LIMITATION OF 6'-0"

Figure 15-2. Height and Length Limitations of Hollow Structural Clay Tile, Both Side and End Construction

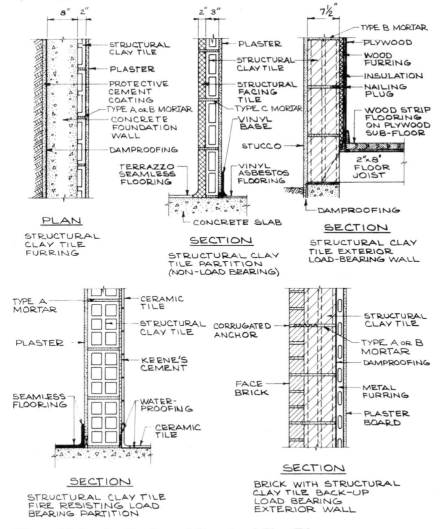

Figure 15-3. Application of Structural Clay Tile

structural clay tile walls as imposed by most building codes are shown in Figure 15-2. The limitations apply to both side and end construction methods.

Mortars. (See also Unit 13, Building with Brick.) Mortars for structural clay tile are type A, used where fireproofing is required, type B, for load-bearing, and type C, for non-load-bearing exterior and partition walls. Mortar joints are generally 1/4-in., struck flush. Figure 15-3 shows typical sections and a plan of structural clay tile, bearing and nonbearing walls.

Water absorption. The water absorption rate of structural clay tile is very important because if it is too great, the water in the mortar will be absorbed by the tile, causing the mortar to set too quickly and thereby weakening the adhesion.

The two grades of structural clay tile, therefore, are controlled as to their water absorption rates in percentages. The LBX tile will absorb 19% and the LB tile 28%. A permissible variation in the rate of absorption of the tile delivered to a construction site should not be more than 12%.

Fire resistance. The fire resistance rating for structural clay tile is based on the number of cells and the wall thickness (see Table 15-2).

Structural Facing Tile

Structural facing tile is used for load-bearing and non-load-bearing walls and partitions and back-up areas where easy maintenance is required and a hard, colored, finish surface is of importance. The two classes of structural facing tile, FTX and FTS, are manufactured with a maximum water absorption rate as indicated in Table 15-3.

The same mortars are used for structural facing tile as are used for structural clay tile except that the joints are generally 1/4 in. The sand used in the mortars should pass a

No. 16 screen. The wall sections in Figure 15-4 show application of structural facing tile.

When structural facing tile is used in construction, such as around a door opening, it is necessary to calculate both horizontal and vertical joints so that there will be a minimum of cutting of the structural facing tile (see Figure 15-5). This is also true for window sills and at door bases where special shapes have to meet both vertical and horizontal dimensions.

Concrete Block

Concrete block varies in weight, color, texture, and coefficients of heat transmission with the type of aggregate used (see Table 15-4).

Effect of different aggregates. The different aggregates in concrete block also greatly affect its strength and denseness but can impart other characteristics which may be desirable. For example, sand and gravel give the concrete block high compressive strength and low water absorption but also denseness and durable strength; cinders decrease compressive strength but increase fire-resistance and insulating properties and also make the blocks nailable; shale, clay, and slag give high strength but also increase fire-resistance and insulating value, besides making the block nailable.

Dimensional limitations. In general, building codes specify height limitations for various thicknesses of concrete block used as supporting walls (load-bearing) such as in residences or single or multistoried buildings (see Figure 15-6).

Height limitations are restricted to 12′-0″ for 8-in. block and 35′-0″ for 12-in. block, and the block must increase in size by 4 in. for every 35′-0″ additional height. Horizontally and vertically, the maximum distance between lateral supports for concrete block walls is 12′-0″ for 8-in. block and 18′-0″

Table 15-3. Maximum Water Absorption (percent) Average for Structural Facing Tile

Class	After Submersion in Cold Water for 24 hrs.	After Submersion in Boiling Water for 1 hr.
FTX	7	9
FTS	13	16

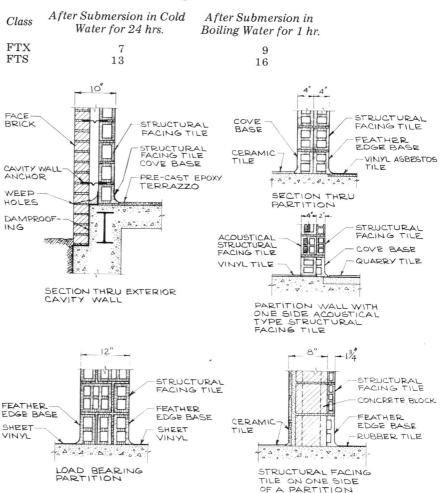

Figure 15-4. Application of Structural Facing Tile

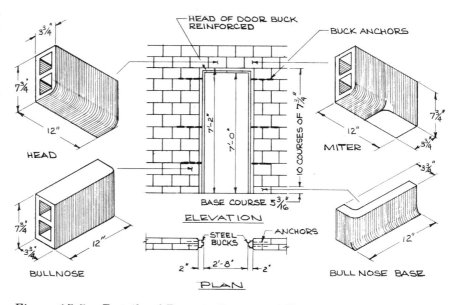

Figure 15-5. Details of Door in Structural Facing Tile Partition

123

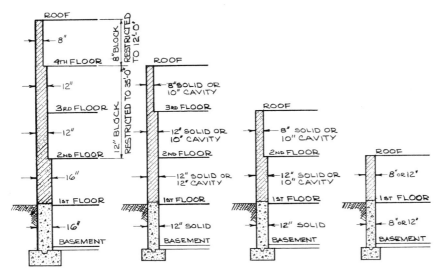

Figure 15-6. General Size Restrictions of Concrete Block When Used as Load-Bearing Walls

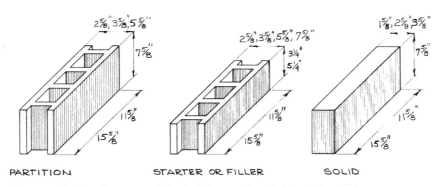

Figure 15-7. Concrete Block Shapes Available Other Than Standard Shapes

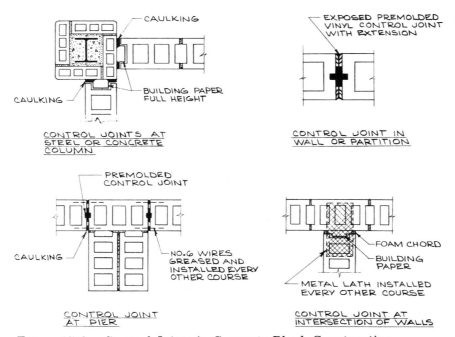

Figure 15-8. Control Joints in Concrete Block Construction

for 12-in. block. Cavity walls are restricted to three-story buildings and have the same restrictions on lateral supports, both vertically and horizontally, as 8-in. block.

Mortars and special shapes. The most frequently used concrete block shapes other than standard (for which see Unit 14) are shown in Figure 15-7. However, manufacturers make other sizes and shapes to meet various special conditions and requirements that occur in construction. The mortars used for concrete block construction are summarized in Table 15-5.

Control joints. In all concrete block construction control joints must be installed to eliminate cracking due to thermal expansion and contraction. In long horizontal walls and partitions, control joints should be installed approximately every 20'-0". Other areas where they should be installed are generally at pilasters, columns, intersections, and openings (see Figures 15-8 and 15-9).

Exterior and interior uses of concrete block. When concrete block is used on exterior walls (except in very dry climates), either the denseness and water resistance of the block or the surface treatment must be considered in order to stop penetration of water into the interior of the building. This can be accomplished by applying a stucco with a waterproof cement, by painting the surface with a waterproof paint, or by using a special dense-surface block (see Figure 15-10, A and B), Figure 15-10, C, shows a load-bearing concrete block partition with open-web steel joists.

Load-bearing and non-load-bearing concrete block interior partitions with various types of surface treatments such as ceramic tile, wood panels, structural facing tile, etc., are shown in Figure 15-11.

Concrete block with special surface finishes. Manufacturers are

Table 15-4. *Effects of Different Aggregates in Concrete Block*

Aggregate type	Weight lbs. per unit	Color	Texture	"U" factor
Sand and gravel	38 to 43	Light gray	Fine	8″ = 0.59 and 12″ = 0.49
Cinder	26 to 33	Dark gray	Coarse to medium	8″ = 0.37 and 12″ = 0.35
Shale, clay, or slag	26 to 33	Light gray	Fine, medium	8″ = 0.33 and 12″ = 0.32

Table 15-5. *Classes of Mortar used with Concrete Block (by volume)*

Class of Mortar	Cementitious Material cu. ft.	Lime Putty cu. ft.	Sand cu. ft.	Type of Support	Exterior or Interior Location
A	1 portland cement	$\frac{1}{3}$	$2\frac{3}{4}$	Heavy loads and below grade	Interior and exterior
A (alternate)	$\frac{1}{2}$ portland cement $\frac{1}{2}$ masonry cement	none	4	Heavy loads and below grade	Interior and exterior
C	1 portland cement	$1\frac{1}{4}$	6	Light loads	Interior
C-1	1 masonry cement	none	3	Light loads	Interior
C-2	1 slag cement	$\frac{5}{8}$	$4\frac{1}{2}$	Light loads	Interior

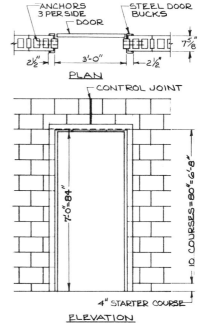

Figure 15-9. *Control Joint at Opening*

researching and producing concrete block with special surface treatments making possible durable, colored, and easily maintained surfaces. These blocks can replace brick, structural facing tile, acoustical materials, etc. At present the generally accepted special surface finished concrete blocks for building construction are acoustical type surfaces and durable surfaces, for interior treatments and also for very dense water-resistant surfaces for exterior walls. The exterior types are identical in size and shape to standard type concrete blocks. The acoustical and special colored, durable types vary in construction details because of the increased dimensions of the surface finishes (see Figures 15-12 and 15-13). The mortars used for these types of concrete block are the same as for standard concrete block. With durable, colored surface finish block, the joints are 1/4-in. flush joints.

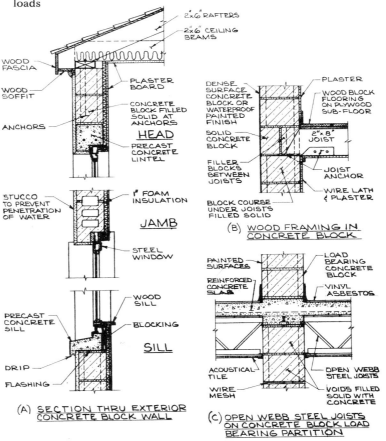

Figure 15-10. *Concrete Block Construction with Waterproof Surface Treatments*

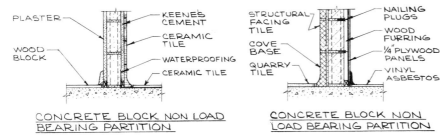

Figure 15-11.   Non-Load Bearing Interior Partitions

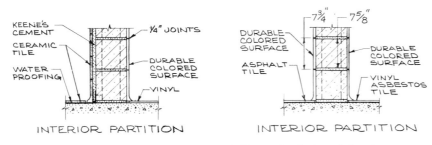

Figure 15-12.   Concrete Block with Durable Colored Surface

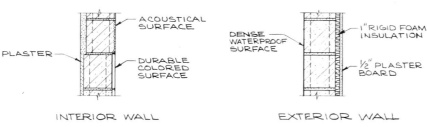

Figure 15-13.   Concrete Block for Interior and Exterior Walls

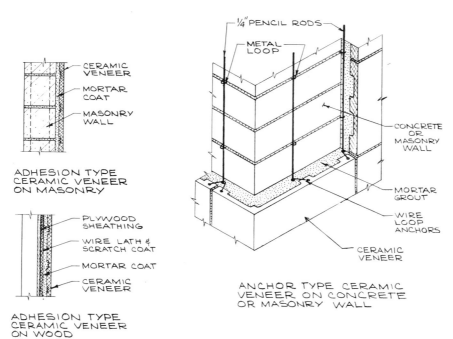

Figure 15-14.   Ceramic Veneer

Terra Cotta (Ceramic Veneer)

When using this material in building construction (see Figure 15-14), it is important to control the possible warpage of the surfaces, because a piece of material this size, when burned in a kiln, can warp to a considerable degree. Necessary restrictions have therefore been established on the acceptable limitations of warpage in a surface. For example, a piece of terra cotta 2-5/8 in. thick, 4 ft., 3 in. high, and 2 ft., 8-1/2 in. wide could warp in burning as much as 2 in. (see Table 15-6). Note that for anchor-type ceramic veneer the minimum number of anchors has also been specified.

Gypsum Block

Gypsum block is used extensively in building construction for interior partitions because of its light weight and good fire-resistant rating. It should never be used in areas where water or heavy moisture are present because it will absorb moisture, become less strong, and will eventually disintegrate. Most building codes limit the height of gypsum block partitions (see Figure 15-15). The faces of gypsum block cannot be used as a finish wall surface and must receive a protecting finish coat, generally of gypsum plaster. When installing a gypsum block partition it is always advisable to install a starter course of concrete block or structural clay tile at the floor.

Table 15-6. *Maximum Face Distortions per Size of Terra Cotta*

Area (in sq. ft.)	Maximum distortion of warpage in inches	Minimum number of anchors
1 sq. ft. or less	$\frac{1}{16}$"	0
1 sq. ft. to 2 sq. ft.	$\frac{1}{8}$"	2
2 sq. ft. to 4 sq. ft.	$\frac{1}{8}$"	3
4 sq. ft. to 12 sq. ft.	$\frac{3}{16}$"	4
12 sq. ft. to 20 sq. ft.	$\frac{3}{16}$" and over	6
Greater than 20 sq. ft.	$\frac{3}{16}$" and over	one for each 35 sq. ft.

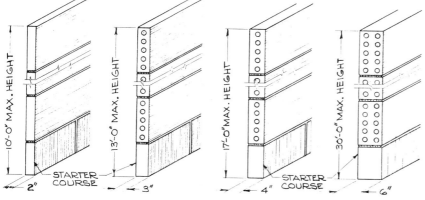

STARTER COURSES TO BE CONCRETE BLOCK OR STRUCTURAL CLAY TILE

Figure 15-15. *Height Limitations of Gypsum Block Partitions*

REVIEW EXAMINATION

1. What are the two grades of structural clay tile?

2. The maximum height that a 6-in. structural clay tile wall or partition can be built is what?

3. Non-load-bearing structural clay tile is installed with type A mortar — true or false?

4. Is the water absorption rate of structural clay tile important?

5. What are the two classes of structural facing tile?

6. The joints for structural facing tile are how big?

7. Can structural facing tile be obtained with both faces having a hard, colored, and easily maintained finish surface?

8. The horizontal and vertical courses should be what, when structural facing tile is to be installed in a building?

9. The type of aggregate used in manufacturing concrete block affects what factors of the concrete block?

10. Do building codes specify height restrictions for thickness of concrete block walls?

11. Do building codes specify height restrictions by the number of stories in a building for thickness of concrete block walls?

12. Control joints are installed in concrete block walls and partitions to stop what?

13. When concrete block is used on exterior walls, what must be done to prevent moisture and water from penetrating into the interior?

14. Concrete block with durable, colored, and easily maintained surfaces are manufactured with this surface on one side only — true or false?

15. Ceramic veneer is manufactured in what two types?

ASSIGNMENT

1. Draw a freehand sketch to scale of a door in a structural facing tile and a concrete block partition, both in plan and elevation.

2. Draw a freehand sketch to scale of control joints in concrete block walls and partitions.

SUPPLEMENTARY INFORMATION

More About Concrete Block

Decorative shapes of concrete block. Concrete block manufactured in shapes with various voids is used to construct pierced decorative walls, partitions, screens, etc. Heights and lengths of the block vary according to the penetrations (see Figure 15-16).

Joints. The joints used for concrete block are shown in Figure 15-17.

Accessories. The various accessories used with other masonry units in order to strengthen, reinforce, anchor, relieve pressure, etc., are shown in **Figure 15-18**.

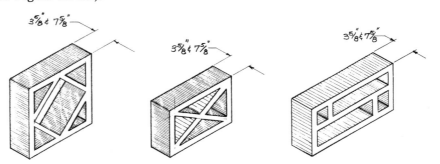

Figure 15-16. Examples of Pierced Concrete Block

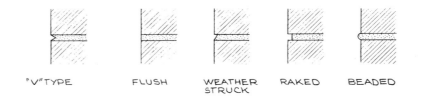

"V" TYPE   FLUSH   WEATHER STRUCK   RAKED   BEADED

Figure 15-17. Mortar Joints Used for Concrete Block

PRE-MOLDED
CONTROL
JOINT FILLER

PRE-MOLDED CONTROL
JOINT FILLER WITH
VINYL TIP

VINYL
FILLED
ROPE

CONTROL JOINT ACCESSORIES FOR CONCRETE BLOCK

REINFORCING

REINFORCING
BLOCK & BRICK

MESH

REINFORCING FOR CONCRETE BLOCK, STRUCTURAL CLAY & FACING TILE

Figure 15-18. Masonry Accessories

16

Stone, Floor, and Wall Tile

INTRODUCTION

Stone and tile are two materials that have been used in building construction from the earliest times. This unit will discuss the kinds of stone and the types of floor and wall tile that are used in present-day building construction.

TECHNICAL INFORMATION

Stone

There are three groups of stone used in building construction: *igneous stone*, formed by solidification from a molten state (for example, granite); *sedimentary stone*, formed by the breaking down of stone into small particles which are cemented together by chemical or organic action, for example, limestone; and *metamorphic stone*, the ultimate product from both igneous and sedimentary stone, formed either by pressure, heat, or moisture or by various combinations of these forces (for example, marble). Table 16-1 summarizes these three groups, giving basic data on the weight and strength of each.

Table 16-1.   Type of Stone, Name of Stone, Weight, and Compressive Strength

Type of Stone	Name of Stone	Weight (lbs. per cu. ft.)	Compressive Strength (lbs. per sq. in.)
Igneous	Granite	162.7 to 184.6	13,000 to 32,000
Sedimentary	Limestone	131 to 170	2,600 to 21,320
	Sandstone	135 to 170	7,700 to 12,000
Metamorphic	Marble	131 to 177.1	12,156 to 22,900
	Slate	175	10,000 to 15,000
	Soapstone	170	9,000 to 14,000

Stone is available for building construction in four categories: (1) dimension (cut) stone; (2) rubble stone (ashlar type); (3) flagstone; and (4) crushed stone. Stone dust and monumental stone are categories of lesser importance, not related to building constructions.

The various types of stonework are identified by the shape and the surface or finish treatment of the stone (see Figure 16-1).

Dimension (cut) stone.   This type of stone is pre-cut to a specific size, squared to dimensions over 24 in. each way, to a specific thickness, and with specific finishes. It is used for exterior veneers, interior wall finishes, flooring, toilet partitions, stair treads, window and door sills, copings, columns, and trim (see Figure 16-2).

131

Dimension (cut) stone is available in rough, smooth, textured,

FIELD STONE

COURSED FIELD STONE

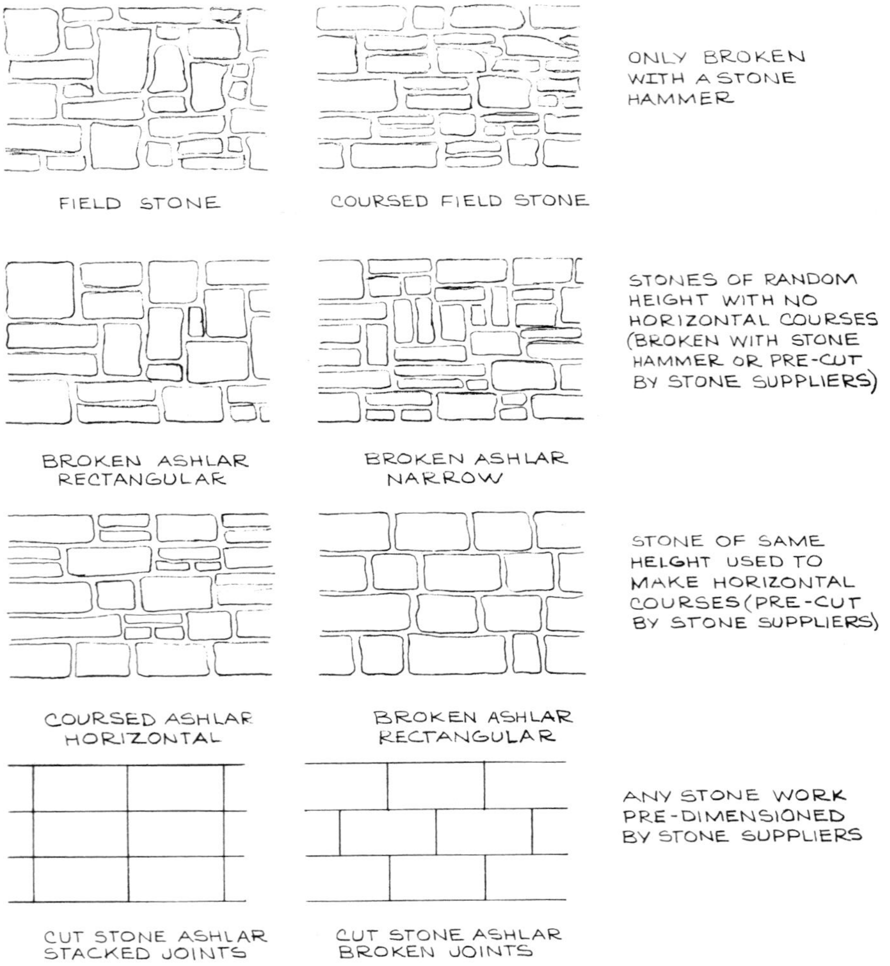

ONLY BROKEN WITH A STONE HAMMER

BROKEN ASHLAR RECTANGULAR

BROKEN ASHLAR NARROW

STONES OF RANDOM HEIGHT WITH NO HORIZONTAL COURSES (BROKEN WITH STONE HAMMER OR PRE-CUT BY STONE SUPPLIERS)

COURSED ASHLAR HORIZONTAL

BROKEN ASHLAR RECTANGULAR

STONE OF SAME HEIGHT USED TO MAKE HORIZONTAL COURSES (PRE-CUT BY STONE SUPPLIERS)

CUT STONE ASHLAR STACKED JOINTS

CUT STONE ASHLAR BROKEN JOINTS

ANY STONE WORK PRE-DIMENSIONED BY STONE SUPPLIERS

Figure 16-1. Stone Work

NON-STAINING FLASHING AND PIN

PITCH

CUT STONE

DRIP

CUT-STONE COPING

CUT STONE

DRIP

NON-STAINING FLASHING

CUT-STONE SILL

Figure 16-2.   Use of Cut Stone

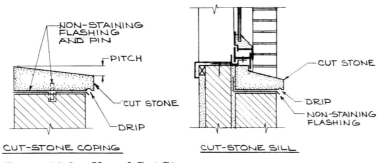

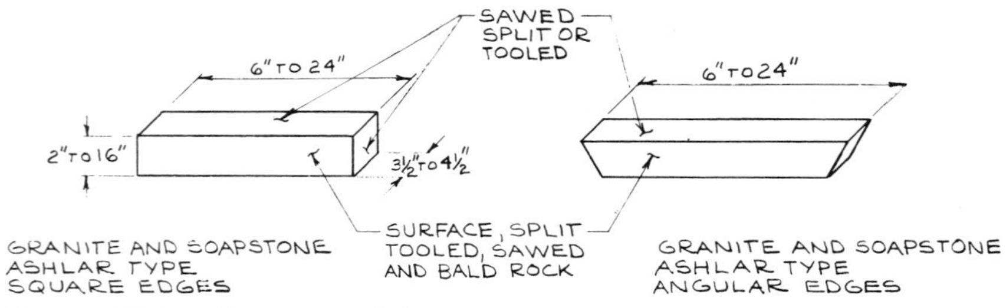

SAWED SPLIT OR TOOLED

6" TO 24"

2" TO 16"

3½" TO 4½"

6" TO 24"

GRANITE AND SOAPSTONE ASHLAR TYPE SQUARE EDGES

SURFACE, SPLIT TOOLED, SAWED AND BALD ROCK

GRANITE AND SOAPSTONE ASHLAR TYPE ANGULAR EDGES

Figure 16-3.   Granite and Soapstone

and polished finishes and in a very wide variety of colors, including numerous variations of veining and color combinations. Stone veneers vary from 7/8 in. thick (minimum) up to 2-1/6 in. thick or more. Generally, the maximum practical size is 8 ft. high and 12 ft. wide; however, larger sizes are now available. All dimension stone veneers are set at a minimum of 1 in. from the back-up material; shop drawings are required to show sizes, joints, anchoring methods, and special shapes. Generally both vertical and horizontal joints are made 1/4 in. The types of stone used for dimension (cut) stone are granite, marble, limestone, slate, and soapstone.

Rubble stone (ashlar type). This type of stone is delivered from the quarries in irregular shapes with certain size limits, usually 12 in. high and 2 ft. long. The stone mills cut this rubble stone into ashlar type veneers, copings, sills, curbing, etc.

The ashlar type veneers vary in shape, size, and finish with the types of stone used. For granite see Figure 16-3 and for limestone see Figure 16-4.

Flagstone. Flagstones are thin slabs of stone, from 7/8 to 2 in. thick, either irregular or squared, with the finished surface available smooth, slightly rough, or polished. They are used for terraces, walks, floors, stair treads, flooring, blackboards, sills, and counter tops (see Figure 16-5).

Crushed stone. Crushed stone consists of granules, chips, or irregular shapes that have been graded and sized. It differs from gravel in that it is generally composed of one type of stone. Sizes start at 1/4 in. and can be obtained in fixed sizes up to 2-1/2 in.

Crushed granite is used for terrazzo-type flooring, artificial stone, pre-cast concrete panels, and surfacing treatments for buildings.

Crushed limestone is used as aggregate for concrete and asphaltic concrete, and as sand for mortar, plaster, and concrete. Screened limestone chips are used for surfacing built-up roofs and in terrazzo, artificial stone, pre-cast concrete panels, and surfacing treatment for buildings.

Crushed marble is used in terrazzo-type flooring, artificial stone, pre-cast concrete panels, surfacing treatment for buildings, and for surfacing built-up roofs.

Stone dust. Waste stone from the quarries, stone mills, and stone suppliers which cannot be used as crushed stone is pulverized into stone dust, which is used as a filler in asphalt-type resilient flooring, shingles, siding, rubber products, paints, paper, linoleum, and textiles.

Monumental stone. Monumental stone is available rough or finished and is used for sculpture, monuments, and the like.

whereas the plastic process yields tile with a hand-made appearance.

All floor and wall tile less than 36 sq. in. are based on the 6-in.

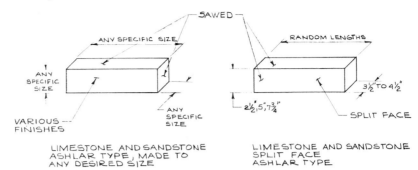

Figure 16-4. Limestone and Sandstone

Figure 16-5. Flagstone

Floor and Wall Tile

Floor and wall tile may be defined as small thin solid slabs, either glazed or unglazed, made from clay or a mixture of clay and other ceramic materials and fired (burned) in kilns to produce a strong, durable material. They differ, as shown in Table 16-2, principally in composition of tile body, surface finish (glazed or unglazed), process of manufacture (dust-press or plastic method), and degree of vitrification (extent to which the tile absorbs water).

There are two methods of manufacturing floor and wall tile: the *dust-press process*, in which the damp mix is shaped with steel dies under heavy pressure, and the *plastic process*, in which the mix has sufficient water added so it is plastic and can then be extruded and cut or molded by hand. The dust-press process gives mechanical dimension precision to the tiles,

Table 16-2. Basic Information for Wall and Floor Tile

Type	Grades	Sizes	Mfg. Process	Surface	Vitrification	Use
Glazed wall tile (interior) White or colored	Standard seconds	From a minimum of $2\frac{1}{8}$" by $2\frac{1}{8}$" to a maximum of 72 sq. in.	Dust-press process	Glazed surface impervious to moisture	Nonvitreous	Interior wall surfaces
Ceramic mosaic tile White or colored	Standard seconds	From a minimum of $\frac{1}{4}$" by $\frac{1}{4}$" to a maximum of 6 sq. in.	Dust-press and plastic processes	Unglazed	Semivitreous, vitreous, and impervious	Interior and exterior floors and walls
Glazed weatherproof tile White or colored	Standard	From a minimum of $\frac{1}{4}$" by $\frac{1}{4}$" to a maximum of 6 sq. in.	Dust-press and plastic processes	Glazed surface impervious to moisture	Semivitreous and vitreous	Interior and exterior floors and walls
Quarry tile	Select standard seconds	From a minimum of $2\frac{3}{4}$" by $2\frac{3}{4}$" to a maximum of 81 sq. in.	Plastic process	Unglazed	Vitreous and impervious	Interior and exterior floors, walks, and terraces
Pavers	Standard seconds	From a minimum of 3" by 3" to a maximum of 36 sq. in.	Dust-press and plastic processes	Unglazed	Vitreous and impervious	Interior and exterior floors, walks, and terraces

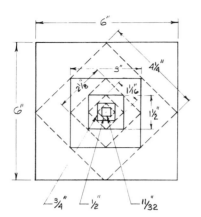

Figure 16-6.   Floor and Wall Tile

square (see Figure 16-6). Through this method all sizes and shapes of floor and wall tile can be combined into almost unlimited patterns and designs.

There are designated grades of tile: glazed wall tile and ceramic mosaics are graded as *standards* and *seconds;* all other types of tile do not have any designated grades. The difference in grades covers warpage and surface defects only, because both grades have the same wear, strength, and sanitary qualities.

Quarry tile. Three grades are available: *select*, *standard*, and *seconds*. Standard grade is controlled as follows: warpage of a maximum of 0.024 in. in the 6-in. length and 0.017 in. in the 4-1/4-in. length; no surface defects visible at a distance of 3'-0". Seconds cannot have a fracture of the body.

Quarry tile is available in 1/2-in., 3/4-in., and 3/8-in. thicknesses.

Sizes range from 2-1/4 to 9 in. in widths, and from 2-3/4 to 9 in. in length.

Pavers. These are unglazed tile having 6 sq. in. or more of facial area, and they are used for floors where heavy traffic is anticipated.

Vitrification. All floor and wall tile are classified into four degrees of vitrification: *non-vitreous*, which can have a moisture absorption of more than 7% of the weight of the tile; *semi-vitreous* in which the limits of moisture absorption is limited to from 3 to 7% of the weight of the tile; *vitreous*, which limits moisture absorption to less than 3% of the weight of the tile; *impervious*, in which moisture absorption is negligible or none. All the different types of floor and wall tile have special trim pieces for bases, caps, corners, moldings, angles, and the like (see Figure 16-7).

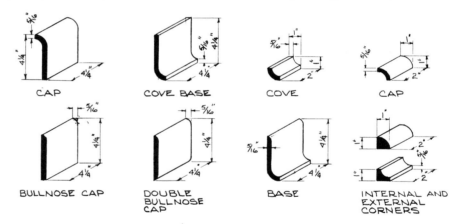

Figure 16-7.   Trim Pieces for Floor and Wall Tile

REVIEW EXAMINATION

1. Name the three groups of stone used in building construction.

2. Name the six types of stone available for building construction.

3. Name two types of metamorphic stone.

4. Name an igneous type of stone.

5. What is the minimum thickness of dimension (cut) stone?

6. Give three uses of crushed stone.

7. The thickness of vertical and horizontal joints for dimension (cut) stone is what?

8. Name the two manufacturing processes used to make wall and floor tile.

9. Floor and wall tile less than 36 sq. in. is based upon what?

10. What are the two grades of glazed wall tile and ceramic mosaic tiles?

11. What are the four degrees of vitrification?

12. What are bases, caps, moldings for wall, and floor tile known as?

ASSIGNMENT

1. Draw a freehand sketch showing the various types of stonework.

2. Draw a freehand sketch showing the various floor layouts for the various shapes of flagstones.

SUPPLEMENTARY INFORMATION

Ceramic Mosaic Tile

Ceramic mosaic tiles are mounted on sheets of paper about 2 sq. ft. in area by the manufacturer. The tiles are spaced to allow for grouting and pointing. After the tiles are installed, the paper is wetted and removed and the tile is ready for pointing. By this method, any design can be factory-mounted on paper so that the design or pattern repeats accurately when the mounted tile on the paper is installed.

Today many glazed wall tiles are made with nibs so that when installed, the joints are all of similar widths (see Figure 16-8).

There are two methods of installing wall and floor tiles: with *cement mortar*, and with *adhesive* (thinset). With the cement mortar method all trim pieces have to have sufficient radius to take care of the thickness of the mortar (see Figure 16-9), whereas with the adhesive method, the trim pieces are the same thickness as the tile (see Figure 16-10).

Color in Stone

Feldspar, hornblende, and mica are responsible for the colors of granite. Feldspar makes granite pink, red, brown, buff, gray, and white. Hornblende and mica make granite dark green and black.

Iron oxides give limestone reddish and yellowish colors and organic materials give limestone gray to black colors. Most limestone for building construction are gray and buff or combinations of these colors.

Marble is available in colors varying from white to black and also with wide variations of veining and color combinations.

Bluestone is available only in a blue-gray color.

Sandstone, which is similar to limestone, is available in white, gray, yellow, brown, and red.

Soapstone is available in gray, green, and blue.

Iron sulfides produce in slate the black, blue, and gray colors, iron oxide the red and purple colors, and chlorite the green color.

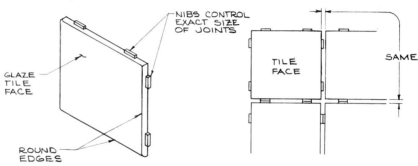

Figure 16-8. Wall Tile

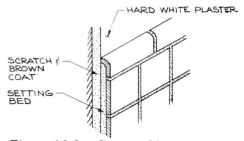

Figure 16-9. Cement Mortar Setting Method

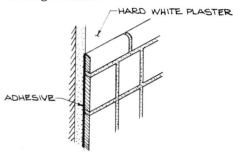

Figure 16-10. Adhesive Setting Method

1. What is one of the principle requirements of concrete?

2. By what is the strength of concrete largely controlled?

3. The slump test measures what in concrete?

4. The largest piece of aggregate should not exceed what fractional part of the concrete slab thickness?

5. A greater strength for concrete can be achieved when its mixing time in minutes is _____ .

6. Air-entraining concrete results when what ingredient is added to the mix?

7. Lightweight concrete is most desirable when used as _____.

8. What is the most important reason for vibrating reinforced concrete?

9. What lumber is well-suited for the building of forms for concrete?

10. What type of form material may be used for two-way ribbed (waffle-type) construction?

11. When depositing concrete into forms, what may cause segregation?

12. The presence of laitance in concrete indicates what fault?

13. In curing, what is the evaporation of water reduced by?

14. What causes volume changes in cured concrete?

15. What is the fundamental cause of shrinkage of concrete at the time of curing?

16. What should exposed construction joints in concrete slabs be filled with?

17. What is necessary when conditions demand the placing of concrete under water?

18. Concrete test specimens should be removed from the molds not sooner than _____ hrs.

19. Concrete is called "reinforced concrete" when what is added?

20. What are prefabricated panels of reinforced concrete extensively used as?

21. What are the soft-mud process, the stiff-mud process, and the dry-press process used to manufacture?

22. The type FBX brick has what distinct quality?

23. Of the grades of brick identified as SW, MW, and NW, what is the minimum compressive strength of MW?

24. What is an important factor in the bonding of brick to the mortar joint?

25. What is adobe brick made of?

26. What is type A mortar intended for?

27. Give the two sizes of mortar joints generally used in brick work.

28. When a new wing is added to an existing building, what must the point of connection be provided with?

29. Name the brick joint that sheds water best?

30. The building that requires an expansion joint must be at least _____ ft. long.

31. What is the coefficient of linear expansion of brick?

32. The filler used in expansion joints is known as a _____ type.

33. Efflorescence, a white stain that appears on brick is caused by what?

34. A masonry unit's cross-sectional area in any plane parallel to the bearing surface is less than which of the following percentage of its cross-sectional area measured in the same plane: 50%, 65%, or 75%?

35. Name three uses of structural clay tile.

36. What is the most widely used masonry unit other than brick?

37. What are the masonry units that are graded LBX and LB?

38. What is the mortar used for structural clay tile where fireproofing is required?

39. What is the maximum height that a 6-in. structural clay tile wall or partition can be built?

40. What is the measurement recommended for structural tile mortar joints?

41. Control joints are installed in concrete block walls and partitions to stop _____ .

42. What is the stone formed by solidification from a molten state?

43. What is the compressive strength of soapstone?

44. How is stone dust used in paints?

45. What does soapstone weigh?

46. What is crushed granite used for?

47. What color does hornblende make granite?

48. Iron oxides give limestone what color?

49. In what colors is soapstone available?

50. Most limestone for buildings is what color?

17

Stone and Ceramic Tile in Construction

INTRODUCTION

This unit covers two topics: types and uses of stone in construction, and the installation of ceramic floor and wall tile. Both were highly developed crafts in former times, and both have been affected by new construction methods, so that here only the basic types and the simple essentials of installation will be presented.

TECHNICAL INFORMATION

Stone

The most generally used categories of stone for building construction are: dimension (cut) stone, rubble stone (ashlar type), and flagstone.

Dimension (cut) stone veneers or thin slabs of granite, limestone, and marble are generally used for exterior veneers, interior wall finishes, flooring, toilet partitions, and stair treads and risers. Moldings, copings, sills, bases, and special trim or shapes that are used with veneers or thin slabs are usually manufactured from the same type of stone as the veneers or thin slabs. Usually all

joints are 1/4 in. and the various methods of joining are shown in Figure 17-1.

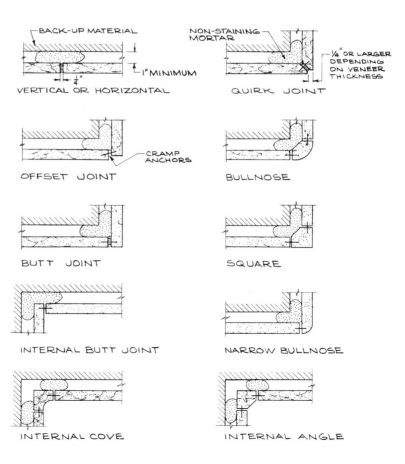

Figure 17-1. *Methods of Joining Stone Veneer*

139

Table 17-1. *Mortars for Setting Stone by Volume*

Mortar	Nonstaining Portland Cement or Nonstaining Waterproof Portland Cement	Sand	Lime
Setting	1	6	1
Pointing	1	2	Sufficient to make a very stiff mortar
Grouting	1	$1\frac{1}{2}$	

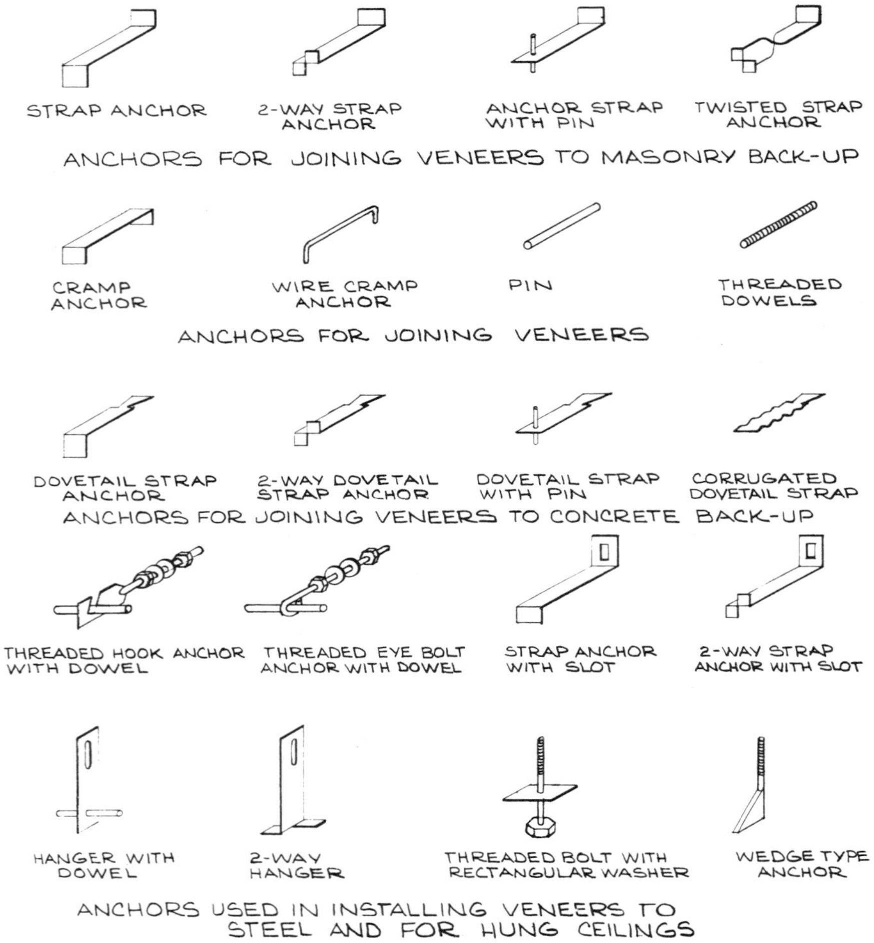

STRAP ANCHOR | 2-WAY STRAP ANCHOR | ANCHOR STRAP WITH PIN | TWISTED STRAP ANCHOR

ANCHORS FOR JOINING VENEERS TO MASONRY BACK-UP

CRAMP ANCHOR | WIRE CRAMP ANCHOR | PIN | THREADED DOWELS

ANCHORS FOR JOINING VENEERS

DOVETAIL STRAP ANCHOR | 2-WAY DOVETAIL STRAP ANCHOR | DOVETAIL STRAP WITH PIN | CORRUGATED DOVETAIL STRAP

ANCHORS FOR JOINING VENEERS TO CONCRETE BACK-UP

THREADED HOOK ANCHOR WITH DOWEL | THREADED EYE BOLT ANCHOR WITH DOWEL | STRAP ANCHOR WITH SLOT | 2-WAY STRAP ANCHOR WITH SLOT

HANGER WITH DOWEL | 2-WAY HANGER | THREADED BOLT WITH RECTANGULAR WASHER | WEDGE TYPE ANCHOR

ANCHORS USED IN INSTALLING VENEERS TO STEEL AND FOR HUNG CEILINGS

Figure 17-2. *Anchors Used in Installing Veneers*

Table 17-2. *Number of Anchors in Relation to Size of Stone Veneer $\frac{7}{8}''$ Thick*

Size of Stone in square feet	Number of Anchors
2	2
2 to 4	3
4 to 12	4
12 to 20	6
20 and greater	1 extra anchor for each additional 3 sq. ft.

Mortars. The mortars used for setting stone, pointing, and grouting are made with nonstaining portland cement or nonstaining waterproof portland cement, sand with no impurities or salt, clean fresh water, and hydrated lime or quicklime putty (see Table 17-1).

In most veneer stonework, the back-up material should be damp-proofed and the space behind the veneer must be open so that no moisture can develop. The veneer is set with spot attachments of mortar. In cases where solid grouting is necessary, nonstaining waterproof portland cement should be used.

Accessories. There are many types of anchors, ties, etc., used in installing veneers that are designed for use with masonry, concrete, and steel (see Figure 17-2).

The number of anchors necessary to support the veneer is controlled by the thickness and the size in square feet (see Table 17-2).

Stone veneer sections. Figures 17-3 and 17-4 show stone veneers with masonry back-up and concrete back-up, including stone wainscot on the interior. Stone veneers are often used for vestibule and entrance ceilings, soffits, and toilet partitions. For methods for supporting stone ceilings and soffits, see Figure 17-5. The installation of stone toilet partitions is shown in Figure 17-6.

Rubble stone (ashlar type). Granite and limestone are used for exterior and interior veneers. Granite (ashlar type) is manufactured in rectangular pieces 3-1/2 to 4-1/2 in. thick, 6 to 24 in. long by 2 to 16 in. high. The edges are sawed, tooled, or split, and squared or angular. The exposed surface is available in various textures, such as sawed, tooled, split, and seamed. When installed, the joints vary from 1/2 to 1-1/2 in. Limestone, on the other hand, is manufactured in two types: (1) sawed on four sides to any specific size, exposed surface available in various finishes; when installed the joints are 1/4 in. for larger sizes and 1/2 in. for smaller sizes; (2) split face ashlar, sawed top and bottom, 3-1/2 to 4-1/2 in. thick, 2-1/4, 5, and 7-3/4 in. high, and random lengths (see Figures 17-7 and 17-8).

Flagstones. These are available in bluestone, granite, limestone, sandstone, slate, and soapstone. Flagstone is 7/8 to 2 in. thick, and is generally used for terraces, floors, walks, stair treads, sills, blackboards, etc.

Bluestone flagstones have square edges, 1 in. thick in $1'-0''$ and $2'-0''$ squares. Granite flagstones have square edges, 7/8 to 1 in. thick, in $2'-0''$ and $3'-0''$ squares. Limestone and sandstone flagstones are 1 to 2 in. thick and are available in $2'-0''$ up to $4'-0''$ squares. Slate flagstones are 7/8 in. thick and are

available in three types: four edges sawed and squared; irregular, with two edges sawed; and random, with no edges sawed. Soapstone flagstones are 7/8 to 1 in. thick, squared edges, and $1'-0''$ up to $3'-0''$ squares.

All flagstones for terraces, walks, and floors are generally installed on a concrete slab base with a mortar bed of 7/8 in. on the interior and 1-1/4 in. on the exterior. The mortar should be a relatively dry mix of 1 part portland cement and 3 parts fine sand. Pointing mortar is 1 part portland cement, 1-1/2 parts fine sand, with enough clean water to make a workable paste (see Figure 17-9). On the exterior, flagstones may be laid in a bed of sand without a concrete slab so that grass can grow in the joints (see Figure 17-10).

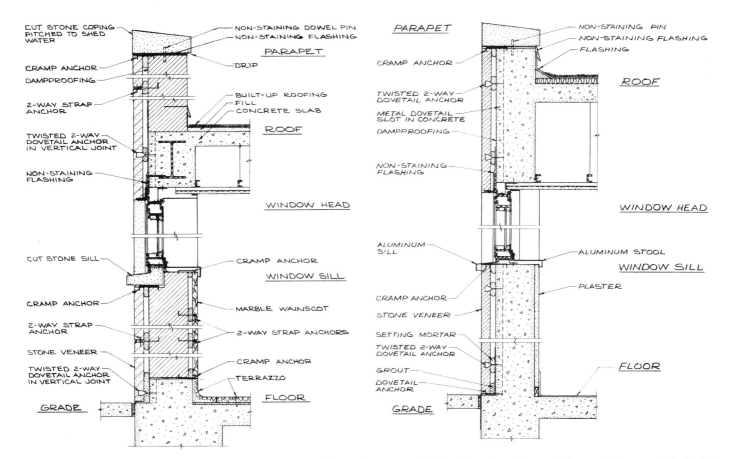

Figure 17-3. Section through Stone Veneer with Steel Construction and Masonry Back-up

Figure 17-4. Section through Stone Veneer with Reinforced Concrete Back-up

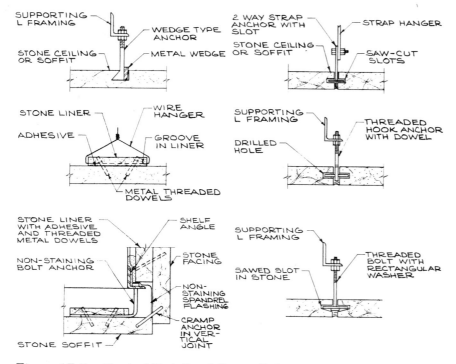

Figure 17-5. Typical Details of Stone Ceilings and Soffits

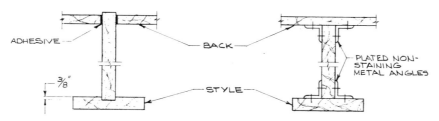

Figure 17-6. Typical Details of Stone Toilet Partitions

Floor and Wall Tile

Terminology. Ceramic tile is a general term used for all types of floor and wall tile. *Glazed tile* is any tile with a glaze on it. *Ceramic mosaic* is small glazed or unglazed tile, mounted on paper. *Mosaic* indicates designs created by artists for floors and walls.

Installation. Floor and wall tile are installed by the cement mortar setting method and the adhesive setting method. For either, the size of joints is controlled to a minimum and a maximum (see Table 17-3). When floor and wall tile are installed with cement mortar, the mixes are controlled by volume for the different coats (see Table 17-4). The sand used for these mortars must be very carefully graded (see Table 17-5).

The installation of floor tile in bathrooms and showers requires that the area be carefully waterproofed (see Figure 17-11). Where excessive water is not present, no waterproofing is necessary. For countertops, particularly in kitchens and bathrooms, the installation of tile at walls, sinks and lavatories, and countertop cooking range require careful detailing (see Figures 17-12 and 17-13).

Figure 17-7. Rubble Stone

Table 17-3. Widths of Joints for Floor and Wall Tile

Size	Type of Tile	Joints Minimum	Joints Maximum
$2\frac{3}{16}$″ square or smaller	Ceramic mosaic glazed or unglazed	$\frac{1}{16}$″	$\frac{1}{8}$″
$2\frac{3}{16}$″ to $4\frac{1}{4}$″	Unglazed tile	$\frac{1}{8}$″	$\frac{1}{4}$″
6″ × 6″ and	Unglazed tile	$\frac{1}{4}$″	$\frac{3}{4}$″
3″ × 3″ and larger	Glazed tile	$\frac{1}{16}$″	$\frac{1}{4}$″
All sizes	Quarry	$\frac{3}{8}$″	$\frac{3}{4}$″

Table 17-4. Mortar Mixes by Volume for Wall and Floor Tile

Mortar	Vertical Surfaces Scratch	Vertical Surfaces Leveling	Vertical Surfaces Setting	Horizontal Surfaces Setting
Portland cement	1 part	1 part	1 part	1 part
Hydrated lime	20%	$1\frac{1}{2}$ parts	$\frac{1}{2}$ to 1 part	10% (maximum)
Sand	4 parts	4 to 7 parts	4 to 7 parts	4 to 7 parts

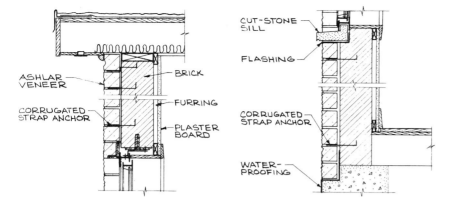

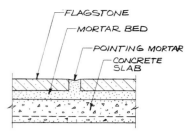

FLAGSTONES ARE PRESSED INTO MORTAR BED TO FINISHED LEVEL AND MORTAR SQUEEZED UP INTO JOINTS IS REMOVED

Figure 17-9. Flagstone in Mortar

Figure 17-8. Ashlar Veneer on Masonry Back-up

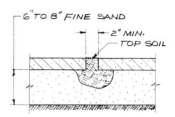

SAND BED RAKED LEVEL AND WETTED, THEN SLATE TAMPED TO DESIRED LEVEL. REMOVE SAND AT JOINTS AND FILL WITH TOP SOIL

Figure 17-10. Flagstone in Sand

Table 17-5. Percentages of Sand Passing Sieve Sizes

Types of Mortar	No. 4	No. 8	No. 16	No. 30	No. 50	No. 100
Setting beds	100%	95% to 100%	60% to 85%	35% to 60%	15% to 30%	0% to 5%
Pointing				100%		0% to 5%

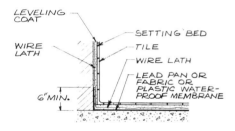

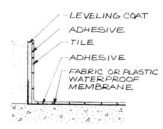

FLOOR TILE WITH WATERPROOFING ADHESIVE METHOD

Figure 17-11. Bathroom and Shower Floor Tile

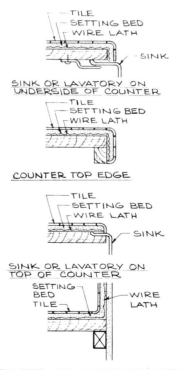

SINK OR LAVATORY ON UNDERSIDE OF COUNTER

COUNTER TOP EDGE

SINK OR LAVATORY ON TOP OF COUNTER

COUNTER TOP INTERSECTING WALL

Figure 17-12. Cement Mortar Method

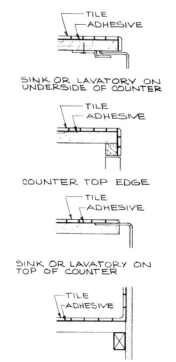

SINK OR LAVATORY ON UNDERSIDE OF COUNTER

COUNTER TOP EDGE

SINK OR LAVATORY ON TOP OF COUNTER

COUNTER TOP INTERSECTING WALL

Figure 17-13. Adhesive Method

143

1. What are the three categories of stone generally used in building construction?

2. Which types of stone are most used for dimension (cut) stone veneers?

3. What types of portland cement are used for setting, pointing, and grouting of dimension (cut) stone veneers?

4. What is the minimum thickness of dimension (cut) stone veneer?

5. Name the four basic types of anchors.

6. What type of stone is used for ashlar type?

7. Name four types of stone from which flagstones are made.

8. What is the general term used to describe all types of floor and wall tile?

9. What are small glazed and unglazed tiles called?

10. What are the two methods used for installing floor and wall tile?

ASSIGNMENT

1. Make a freehand sketch section to scale showing the use of dimension (cut) stone veneer with masonry back-up through exterior wall of a one-story building.

2. Make freehand sketches showing installation by adhesive method of ceramic mosaic tile on a kitchen countertop with sink.

SUPPLEMENTARY INFORMATION

Limestone Surface Texture

The surface texture or fineness of grain of limestone for building construction is graded A — statuary, B — select, C — standard, D — rustic, E — variegated, and F — Old Gothic. Grades A, B, C, and D vary only in fineness of grain from fine to coarse and are buff or gray in color. Grade E is unselected grain size and the color is a mixture of gray and buff. Grade F is a mixture of grades D and E but includes stones with seams and markings.

Marble Classifications

Marbles are classified on the basis of their natural qualities. Grade A is for sound marbles with uniform and good working qualities. Grade B is similar to A except that they have fewer good working proper-

ties. Grade C marbles have uncertain working qualities and contain flaws, voids, veins, and lines of separation. Grade D is similar to C but has poor working qualities and larger proportion of flaws, voids, etc.

Protection of Stone

All types of stonework, when installed, should be protected from damage from construction traffic and also from possible staining by other materials during construction.

Tile Accessories

Manufacturers supply in tile and in matching colors of tile, bathroom accessories such as soap dishes, towel bars, toilet paper holders, etc. The accessories are made in tile sizes so they can be easily installed. Recessed types should be located on plans and details so that openings and rough boxes can be installed in the back-up material to receive them.

Tile Cleavage Plane

When installing large floor areas with tile, it is necessary to install a cleavage plane because the expansion and contraction of the structural slab may cause cracks in the setting bed and tile joints. A layer of bitumen-saturated felt building paper, 13.9 lbs. per 100 sq. ft., is laid directly on the concrete slab, then wire mesh and the setting bed are installed. By this method this cleavage plane separates the slab from the tile and any movement of the slab is independent of the tile.

Epoxies for Exterior Stone Veneers

Because of the epoxy adhesives and new anchoring systems, marble, granite, and limestone veneers and pre-cast panels with stone veneers are finding greater application in exterior facing for buildings. The veneers 7/8 in. thick and greater are applied with a grid system (see Figure 17-14).

With epoxies, stainless steel U-shaped clips are bonded with epoxy adhesives to the stone veneer and then bolted with clip angles to the structural building frame (see Figure 17-15).

Veneers with small angles plus epoxy bonding are able to form entire column covers of almost any shape or form (see Figure 17-16).

The pre-cast concrete panels with stone veneers have the advantage that the back-up material onto which they are installed can have any type of clip, anchor, or other fastening device cast into or applied onto the back-up material to support the panels (see Figure 17-17).

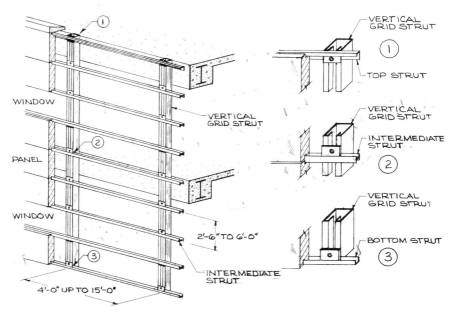

Figure 17-14. *Application of Veneer with Grid System*

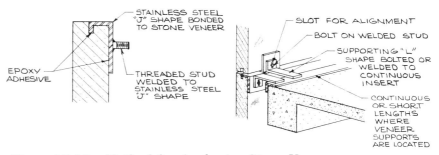

Figure 17-15. *Method for Anchoring Stone Veneer*

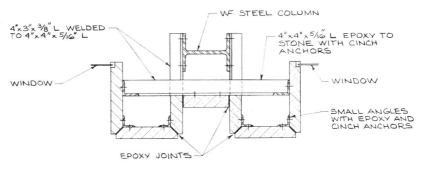

Figure 17-16. *Vertical Column Cover*

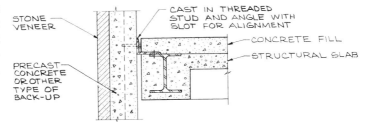

Figure 17-17. *Precast Concrete Panel with Supporting Clips*

18

Wood

INTRODUCTION

Wood, one of our earliest building materials, may still be called a contemporary material because of the many new uses to which it has been put in recent years. Through improved methods of cutting and milling, thin veneers can be produced which, when glued together in layers, form a stronger board than that which is cut from the tree. Wood chips and wood pulp bonded with adhesives into panels and acoustical tile, and small pieces of solid wood laminated to make structural beams, girders, columns, and arches are some of the contemporary uses of wood. Also, a whole new group of surfacing woods and wood products with transparent plastics which give permanent finished surfaces has been developed and greatly improved in recent years. These are some of the items that will be further dealt with in the following units. But in this unit, some of the fundamental facts about wood will be reviewed. Types of wood, seasoning, grading, sizes, quality standards, and classifications will be emphasized.

TECHNICAL INFORMATION

Softwoods and Hardwoods

It is generally known that all lumber is divided into two major groups; namely, the softwoods and the hardwoods. The softwoods, in general, are from coniferous (evergreen) trees, while the hardwoods are from deciduous leaf-bearing trees (those which drop their leaves at the end of the season).

The words "softwoods" and "hardwoods" are used mainly as a matter of custom, for not all so-called softwoods are soft, nor are all so-called hardwoods necessarily hard. For example, Longleaf Southern Yellow Pine and Douglas Fir, both belonging to the softwood family, are harder than poplar and basswood, which belong to the

Table 18-1. Hardness to Softness of Deciduous and Coniferous Tree

Type of Wood	Soft	Relatively Soft	Hard
Hardwoods (deciduous) leaf-bearing	Alder, Basswood, and Poplar	Magnolia and Sweetgum	Walnut, Birch, Cherry, Oak, and Maple
Softwoods (coniferous) evergreens	White Pine, Ponderosa Pine, and Cedar	Cypress, Redwood, Spruce, Fir, and Hemlock	Douglas Fir and Longleaf Southern Yellow Pine

147

hardwood family (see Figure 18-1 and Table 18-1).

Seasoning. All lumber when first cut from trees is called "green" wood. Before it can be used, it should be seasoned either by air-drying or kiln-drying, which will reduce the moisture content. Seasoning of lumber will reduce its weight, reduce shrinkage and warping, and increase the strength of the lumber, its nail-holding power, and its ability to hold paint.

Shrinkage, distortion, and warping in lumber are caused by the location of the annual growth rings of a tree. Therefore, how lumber is cut from a tree is of utmost importance, as Figure 18-2 clearly shows. Some boards, such as A, will warp, and others, such as B, will remain flat; post C will remain square, but post D will be distorted in shape.

Grading of lumber for use. The grade of a piece of lumber is established by the number, character, and location of imperfections that will lower the strength, weaken the durability, and affect the use of the lumber. The most common imperfections are knots, checks, and pitch pockets.

American Lumber Standards Association specifies that lumber be classified according to its principal uses.

1. *Yard lumber* — primarily that which is used in light frame construction.

2. *Structural lumber* — that which is 5 in. or more in thickness and width, and is graded on the basis of strength.

3. *Factory* or *shop lumber* — that which is intended for further manufacture and is graded on a basis of the proportion of area of a piece of lumber for the production of a limited number of cuttings of specified minimum size and quality.

Sizes of lumber. Commonly used softwood boards and dimension lumber established by the American Lumber Standards are given in Table 18-2.

Yard Lumber

This is lumber for general building purposes, including house construction (see Figure 18-3). It is less than 5 in. in thickness and is sold in classifications by dimensions as follows:

1. *Strips* — less than 2 in. thick, not over 8 in. wide.

2. *Boards* — less than 2 in. thick and 8 in. or more in width.

3. *Dimension lumber* — all yard lum-

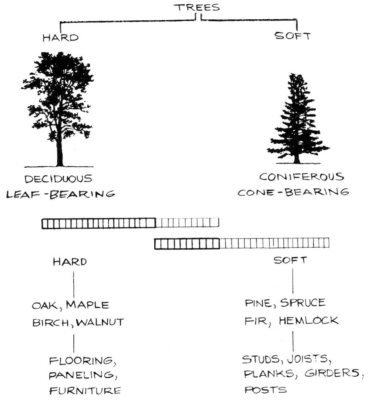

Figure 18-1. Soft and Hard Wood Used for Building

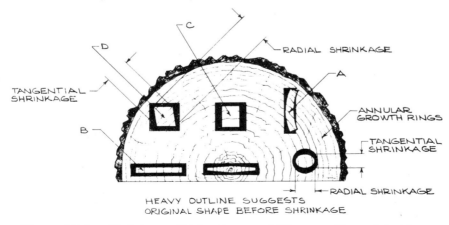

Figure 18-2. Shrinkage, Distortion, and Warpage Caused by Annular Growth Rings

ber with the exception of boards, strips, and timbers. Specifically, it is yard lumber between 2 and 5 in. thick and 2 in. or more in width.

4. *Planks* — yard lumber that is 2 in. to less than 4 in. in thickness and 8 in. or over in width.

5. *Scantlings* — yard lumber that is 2 in. to under 5 in. in thickness and less than 8 in. in width.

6. *Heavy joists* — yard lumber 4 in. to less than 6 in. in thickness and 8 in. or over in width.

Quality standards of yard lumber. Yard lumber is classified into two main divisions on the basis of quality, namely:

1. *Select lumber* — that which is generally clear, contains no defects or only a limited number with respect to both size and type, and is smoothly finished and suitable for use as a whole for finishing purposes or where large clear pieces are required. It is further classified by grades A, B, C, and D. Grades A and B are suitable for natural finish; Grades C and D have defects and blemishes of somewhat greater extent than A and B, which can be covered by paint.

2. *Common lumber* — this is classified by the American Lumber Standards Association as lumber having defects and blemishes not suitable for finishing purposes, but quite satisfactory for general utility and construction purposes. Lumber under this heading is also further classified into Numbers 1, 2, 3, 4, and 5 common. In Numbers 1 and 2 common, all defects and blemishes must be sound. Such lumber is used for purposes in which surface covering or strength is required. Numbers 3, 4, and 5 common contain very coarse defects that will cause some waste within a piece. Such lumber is used where strength is not required.

Structural Lumber

Structural timbers are primarily classified according to size and use as follows (see also Figure 18-4):

Beams and *Stringers* — lumber 5 in. or more in thickness and 8 in. or more in width, graded with respect to its strength in bending when loaded on the narrow face.

Joists and *Planks* — lumber 2 in. to slightly less than 5 in. thick, and 4 or more in. in width, graded with respect to its strength in bending when loaded either on the narrow face or on the wide face as a plank.

Table 18-2.   Softwood Lumber Sizes Government's Acceptance of New Lumber Standards

Nominal Size (inches)	Size when Dry or Seasoned, 19% Moisture Content or Under (inches)	Size when Green or Unseasoned, Over 19% Moisture Content (inches)
1 × 4	$\frac{3}{4} \times 3\frac{1}{2}$	$\frac{25}{32} \times 3\frac{9}{16}$
1 × 6	$\frac{3}{4} \times 5\frac{1}{2}$	$\frac{25}{32} \times 5\frac{5}{8}$
1 × 8	$\frac{3}{4} \times 7\frac{1}{4}$	$\frac{25}{32} \times 7\frac{1}{2}$
1 × 10	$\frac{3}{4} \times 9\frac{1}{4}$	$\frac{25}{32} \times 9\frac{1}{2}$
1 × 12	$\frac{3}{4} \times 11\frac{1}{4}$	$\frac{25}{32} \times 11\frac{1}{2}$
2 × 4	$1\frac{1}{2} \times 3\frac{1}{2}$	$1\frac{9}{16} \times 3\frac{9}{16}$
2 × 6	$1\frac{1}{2} \times 5\frac{1}{2}$	$1\frac{9}{16} \times 5\frac{5}{8}$
2 × 8	$1\frac{1}{2} \times 7\frac{1}{4}$	$1\frac{9}{16} \times 7\frac{1}{2}$
2 × 10	$1\frac{1}{2} \times 9\frac{1}{4}$	$1\frac{9}{16} \times 9\frac{1}{2}$
2 × 12	$1\frac{1}{2} \times 11\frac{1}{4}$	$1\frac{9}{16} \times 11\frac{1}{2}$
4 × 4	$3\frac{1}{2} \times 3\frac{1}{2}$	$3\frac{9}{16} \times 3\frac{9}{16}$
4 × 6	$3\frac{1}{2} \times 5\frac{1}{2}$	$3\frac{9}{16} \times 5\frac{5}{8}$
4 × 8	$3\frac{1}{2} \times 7\frac{1}{4}$	$3\frac{9}{16} \times 7\frac{1}{2}$
4 × 10	$3\frac{1}{2} \times 9\frac{1}{4}$	$3\frac{9}{16} \times 9\frac{1}{2}$
4 × 12	$3\frac{1}{2} \times 11\frac{1}{4}$	$3\frac{9}{16} \times 11\frac{1}{2}$

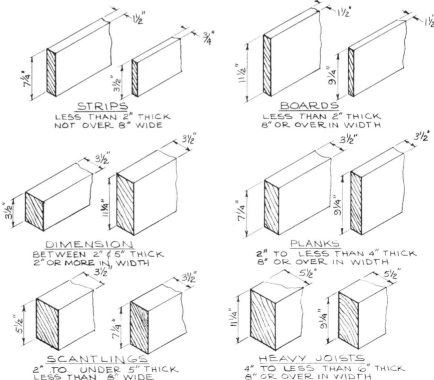

Figure 18-3.   Yard Lumber for General Building Purposes

Posts and *Timbers* — lumber that is square or approximately square in cross section, 4 inches and larger, graded for use as posts or columns carrying longitudinal load, but adopted for uses in which strength in bending is not important.

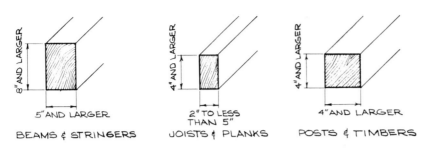

Figure 18-4. Structural Lumber

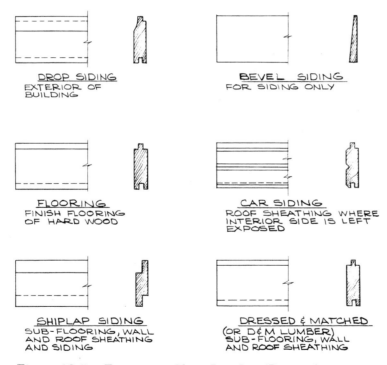

Figure 18-5. Factory or Shop Lumber (Pattern)

Grading by strength. All structural lumber is graded according to its strength in bending. The different species of softwood structural lumber are further tabulated by allowable unit stresses. For example, 2-in. thick Southern Pine (dense, structural, 86% kiln-dried) has a fiber stress in bending (F) and a tensile strength parallel to the grain (T) of 3000 PSI, whereas 2-in. thick Redwood (dense structure) has only 1700 PSI in fiber stress in bending and in tensile strength parallel to the grain.

Factory or Shop Lumber

This is a special-purpose lumber machined from common boards. It is milled in various patterns, the most common of which are those used as subflooring, siding, wall and roof sheathing, finished flooring, and car-siding (see Figure 18-5).

Ship-lap — this lumber is used for subflooring, wall and roof sheathing and siding; it has lapped edge and makes a fairly tight joint between boards, but not as tight as dressed and matched lumber.

Dressed and Matched (D & M) Lumber — this is primarily used for subflooring, wall and roof sheathing, or where a tight joint is needed between boards. Edges are of tongue-and-groove design.

Flooring — this refers to lumber used for finished flooring. It is of hardwood and the harder species of softwoods, cut in thicknesses of 5/16 and 25/32 in.

Siding — several profiles of siding are available in softwoods, varying according to local popularity and design requirements.

Car-siding — has, in addition to the tongue-and-groove design, edges that are beveled and a beveled groove on one surface. This is also called V-joint siding. It is used for roof sheathing where the interior face is left exposed, and for vertical siding.

Drop-siding — used primarily for the exterior of buildings, this is provided with tongue-and-groove edges and may also have ship-lap edges.

Wedged Siding — this is a plain tapered piece of wood.

1. Name four of the hardwood trees.

2. Name four of the softwood trees.

3. Yard lumber is primarily used for what purpose?

4. Name the three lumber classifications specified by the American Lumber Standards Association.

5. What is factory or shop lumber? Give an example.

6. Give the definition of a board. What are its dimensional limitations?

7. What is the actual size of a 2 x 8?

8. Is a 2 x 8 a plank, a board, or is it dimension lumber?

9. What are the two divisions of yard lumber in respect to quality standards?

10. Select lumber has how many grade classifications?

11. Common lumber has how many grade classifications?

12. Structural lumber is classified according to what two factors?

ASSIGNMENT

1. Sketch four types of factory or shop lumber.

2. Sketch four types of yard lumber.

3. Sketch two types of structural lumber.

SUPPLEMENTARY INFORMATION

Moisture Content

Wood swells as it absorbs moisture and shrinks as it loses moisture. In wood there are two types of water: *free water* in the cell cavities and intercellular spaces, and *absorbed water* held in the fibers and rays. When all free water is removed but the absorbed water remains, the fiber saturation point is reached. This is approximately 30% of total moisture content for all species of trees. Shrinkage occurs below the fiber saturation point, and at 15% moisture content about one-half of the total possible shrinkage is attained. For each following 1% loss or increase in moisture below or above the fiber saturation point, the wood shrinks or swells 1/30 of the total shrinkage or swelling. Wood should therefore be seasoned, and air-dried or kiln-dried to a specified moisture content adaptable to the various regions of the United States. The map shown in Figure 18-6 indicates the areas in the United States where wood is specified for its moisture content.

Wood Preservatives

Wood is attacked by fungi, insects, and marine borers. Decay in wood is caused by fungi; most decay occurs where the moisture content is above the fiber saturation point. Insects, most commonly termites, both subterranean and nonsubterranean, attack wood because it is their food. Carpenter ants and powder-pest beetles also attack wood, but to find shelter, not food. Marine organisms attack

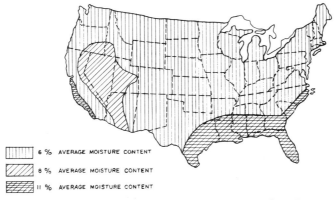

6 % AVERAGE MOISTURE CONTENT

8 % AVERAGE MOISTURE CONTENT

11 % AVERAGE MOISTURE CONTENT

Figure 18-6. Average Moisture Content for Interior Finish Woodwork in the United States

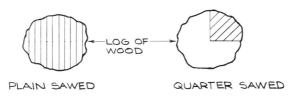

PLAIN SAWED   QUARTER SAWED

Figure 18-7. Types of Lumber Sawing

wood placed in salt water or brackish water.

There are two types of wood preservatives that will protect wood from fungi, insects, and marine organisms:

1. Oil-type, such as creosote.

2. Water-borne salts, such as chromated zinc chloride.

Types of Lumber Sawing

A log of wood is sawed by two methods: *plain-sawing* and *quarter-sawing*, as shown in Figure 18-7 and Table 18-3.

Wood Grade Markings

Typical grade stamp markings that may be found on boards in local lumber yards are indicated in Figure 18-8.

Table 18-3. Advantages of Plain-sawed and Quarter-sawed Lumber

Plain-Sawed	Quarter-Sawed
Less waste	Ray and grain patterns pronounced
Affects strength less	Less raised grain, surface checks, and splits
Shrinks and swells less in thickness	Shrinks and swells less in width
Costs less	Wears evenly

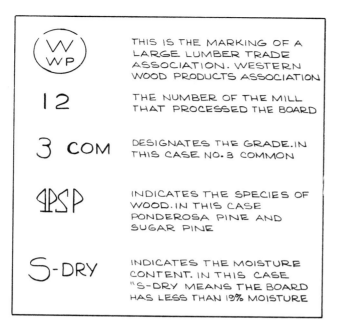

Figure 18-8.   Some Typical Wood Grade Markings

19

Wood and Wood Products

INTRODUCTION

This unit deals with building products that are made from wood. Although there is an endless variety of wood products, some of the more important are considered in this unit. Plywood and its many uses, particle board, wood fiber, and pulp products and their numerous applications are discussed, together with information on jointing and edge treatments.

TECHNICAL INFORMATION

Wood may be sliced into thin veneers; it may also be processed mechanically and chemically into pulp, chips, and fibers, and the waste from wood such as sawdust, shavings, and chips may be and are used to manufacture a great variety of wood products.

Plywood

Plywood is made of softwood used in building construction and hardwood used for cabinet work and furniture. It is made of an odd number of layers or plies of thin wood veneer bonded together so

that the direction of the grain of adjacent plies is at right angles. The outside plies are called "faces" and

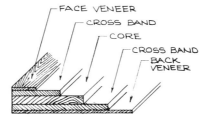

5-PLY VENEER CORE

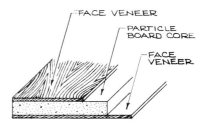

3-PLY PARTICLE BOARD CORE

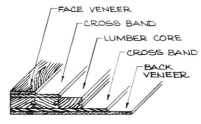

5-PLY LUMBER CORE

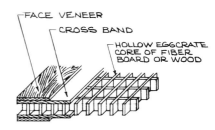

5-PLY HOLLOW CORE

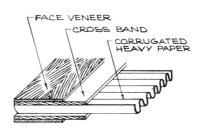

5-PLY CORRUGATED CORE

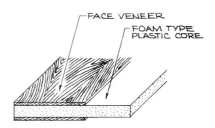

3-PLY FOAM PLASTIC CORE

Figure 19-1. Plywood

155

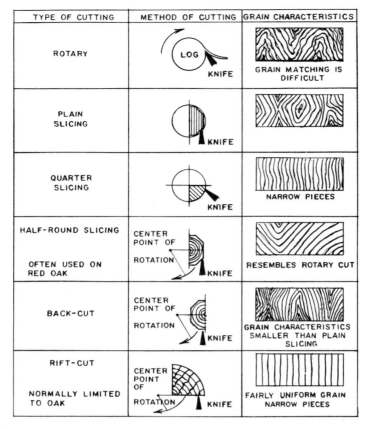

TYPE OF CUTTING	METHOD OF CUTTING	GRAIN CHARACTERISTICS
ROTARY	LOG KNIFE	GRAIN MATCHING IS DIFFICULT
PLAIN SLICING	KNIFE	
QUARTER SLICING	KNIFE	NARROW PIECES
HALF-ROUND SLICING OFTEN USED ON RED OAK	CENTER POINT OF ROTATION KNIFE	RESEMBLES ROTARY CUT
BACK-CUT	CENTER POINT OF ROTATION KNIFE	GRAIN CHARACTERISTICS SMALLER THAN PLAIN SLICING
RIFT-CUT NORMALLY LIMITED TO OAK	CENTER POINT OF ROTATION KNIFE	FAIRLY UNIFORM GRAIN NARROW PIECES

Figure 19-2. The Six Principal Methods of Cutting Veneers

Table 19-1.   Grades of Face Treatment

Grade	Face or Outer Treatment
A	Smooth surface suitable for fine finish or painting
B	Somewhat lower in appearance quality than A
C and D	Have some defects — used where appearance is not important
N	Natural finish, free of all open defects

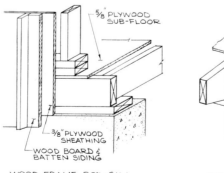

WOOD FRAME BOX-SILL CONSTRUCTION WITH PLYWOOD SHEATHING AND SUB-FLOORING

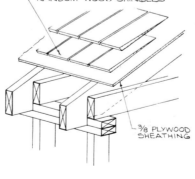

WOOD FRAME ROOF CONSTRUCTION WITH PLYWOOD ROOF SHEATHING AND WOOD SHINGLES

Figure 19-3. Plywood Wall and Roof Sheathing with Board and Batten Siding and Wood Roof Shingles

"backs," while the center ply is called the "core." In the case of five or more plies, the inner plies that are bonded directly and at right angle to the face and core are called the "cross-bands" (see Figure 19-1).

Plywood panels are made with cores of a wood veneer, particle board, solid wood, wood fiber and pulp products, and other types of materials. In the veneer core panel, all plies are veneer less than 1/4 in. thick. Panel thicknesses generally range from 1/8 to 1-1/8 in., consisting of an odd number of plies ranging from 3 to 11 or more. Almost all softwood panels have veneer cores.

Particle board cores are medium-density board made from wood particles. They are being used more and more, replacing the lumber core panel and the veneer core in plywood because of their dimensional stability.

The solid lumber core is composed of strips of lumber edge-glued into a solid slab. This type is usually 5-ply, 3/4 in. thick; for special uses, thicknesses from 1/2 to 1-1/8 in. are manufactured (see Figure 19-1).

Plywood veneer cuts. The manner in which veneers are cut determines the various visual effects obtained. Two woods of the same species with their veneers cut differently will have entirely different visual characters, even though their colors are similar.

Six different methods of cutting plywood veneers are used depending on the type of veneer required. For example, the log is placed centrally in a lathe and turned against a razor-sharp blade (like unwinding a roll of paper) to produce what is known as the "rotary" cut. Other methods of cutting and slicing (see Figure 19-2) are plain slicing, quarter slicing, half-round slicing, back cut, and rift cut. The different cuts produce different grain characteristics.

Basic types of plywood. Plywood is manufactured in two basic types; namely, *exterior* and *interior*. Adhesives for bonding exterior-type panels are waterproof. Glues used for interior plywood are lightly moisture-resistant but not waterproof.

Grades of face treatments. Plywood panels are generally available in five different face or outer veneer treatments (see Table 19-1).

Available sizes of plywood panels. Standard sizes of plywood panels are 4 by 8 ft., but can also be furnished in widths of 2-1/2, 3, and 3-1/2 ft., and in lengths of 5, 6, 7, 9, 10, and 12 ft. Larger sizes such as 16 by 24 ft., or even 50 ft. in length are manufactured for specific uses.

Plywood sizes for exterior uses. When plywood is used as outside siding, a minimum thickness of 3/8 in. is required. When used as wall sheathing (see Figure 19-3), a thickness of not less than 5/16 in. is acceptable when the stud spacing is 16-in. OC, and 3/8 in. is permissible when the stud spacing is 24-in. OC. For roof sheathing, a 5/16-in. may be used with 16-in. OC and 3/8 in. for rafters, 24-in. OC.

Generally, when plywood is used for wall or roof sheathing, it is not necessarily the strength of the plywood that controls the thickness, but the holding power of the nails required for the type of roofing or siding selected.

Jointing of exterior plywood panels. Jointing of plywood panels must be carefully considered if water is to be prevented from entering the building. Horizontal joints may be shiplapped by fitting the ends of two panels or by introducing a molding that may also serve as a design treatment. Shiplapping may be used on all four edges of the panel to form a weathertight joint. With adjacent panels, Figure 19-4 illustrates some of the horizontal and vertical jointing of

plywood panels used for exterior sheathing and siding.

Jointing of interior plywood panels. The jointing of plywood when used for interior finishes and cabinets and built-in furniture depends on design considerations and the handling of the edges of the plywood (see Figure 19-5).

Fiber and Wood Pulp Products

Logs of wood are converted into wood pulp, chips, and fibers by mechanical or chemical processes. These converted materials are then processed into various types of papers, boards, and insulating materials (see Table 19-2).

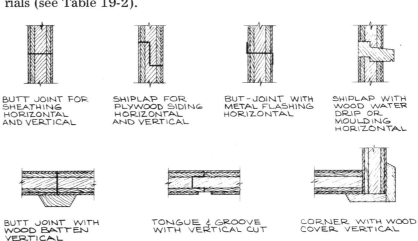

Figure 19-4. Exterior Vertical and Horizontal Joining of Plywood Sheathing and Siding

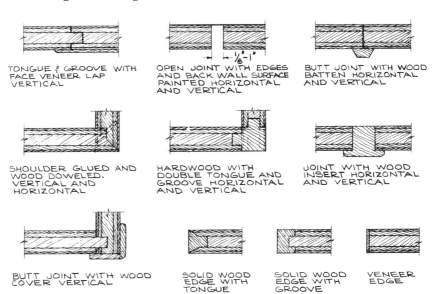

Figure 19-5. Interior Plywood Joining and End Grain Treatments

Fiberboard. Fiberboards are manufactured from wood pulp, waste paper, and fibers. They are lightweight, noncombustible, strong, and possess good acoustical and insulating values. They are available with various surface treatments and find a variety of uses in the construction field for sheathing, insulation, and as interior and exterior finishes for residential buildings.

Hardboard. Wood chips of controlled size are placed in a vat under high-pressure steam. When this pressure is released, the chips blow apart and the fibers and the wood lignin are separated from the unusable materials. This mixture is formed into boards, which are finally compressed with heat to form hard, dense boards.

There are three types of hardboard: *standard*, *tempered*, and *low-density*.

Standard and low-density boards are used only in the interior of buildings, whereas tempered boards are used on the exterior and interior of buildings and in interior areas where moisture and water are present.

Tempered boards are standard hardboard impregnated with oils and resins and subjected to high heat; it is available with a variety of special surface finishes such as grooves, textures, and black color.

Particle board. Particle board (chip board) is a relatively new product which is dry-formed of particles of wood bonded together with synthetic resin or other binders. It might be considered 3-ply board because its outer layers are made of finely cut flakes to provide a smooth, hard surface, and the inner layer is made of more coarsely cut material. In addition to its many uses, the particle or chip board is primarily used as a core stock for high-pressure plastic laminate veneers and hardwood veneers. As core stock, it is used by manufacturers of furniture, cabinets, countertops, wall paneling, partitions, and doors of all types. Figure 19-6 illustrates some of the details in the use of the particle board for cabinet doors, millwork, and sliding closet doors.

Table 19-2. Commonly Used Fiber and Wood Pulp Products

Material	Shape	Use
1. Acoustical	1/2" to 2" thick, in squares or rectangles, in 12" modules, maximum 24" × 48"	Acoustical treatment for walls and ceilings
2. Building paper (sheathing paper)	In rolls, asphalt-impregnated or coated, weighing from 3 to 10 lbs. per 100 sq. ft.	Sheathing paper, vapor barriers, deafening paper under wood floors, and temporary protective covering of finish materials
3. Fiberboards	Boards 3/8" to 1-7/8" in thickness, 4'-0" wide and 8'-0" long; in 1/2" thickness it is available in panels 8'-0" in width and 14'-0" in length	Interior walls and ceilings, exterior sheathing, and finish exterior walls in residential work
4. Flashing	In rolls, laminated with 1-, 2-, or 3-oz. copper foil	Concealed flashing, waterproofing, and vapor barrier
5. Hardboards	Tempered or standard, 1/8" to 5/16" in thickness, 4'-0" wide, and 4'-0" to 12'-0" in 2'-0" increments and 16'-0" in length	*Standard* is used for interior work such as underlayment, cabinet, and millwork; *tempered* is used for both interior and exterior work such as siding, cabinet and millwork, and panelling where one surface has been given a special surface finish
6. Particle board (chip board)	Boards 1/4" to 3/4" and thicker, 4'-0" wide and 8'-0" to 16'-0" long	Cores for wood and plastic laminate veneers, cabinet and millwork
7. Roofing felt	In rolls weighing from 12 to 30 lbs. per 100 sq. ft.	Built-up roofing, dampproofing, and waterproofing,

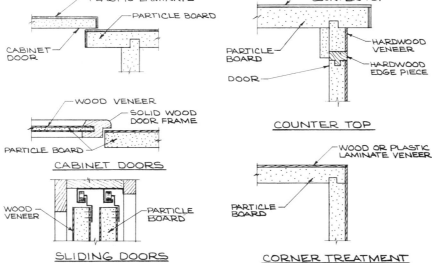

Figure 19-6. Details of Particle Board Used as Cabinet Doors, Millworks, and Sliding Doors

1. What are the outside plies of a plywood panel called?

2. What are the plies to which the face and back are glued in a 5-ply panel called?

3. Name the three types of plywood cores.

4. What are the two basic types in which plywood is manufactured?

5. What is the common size of plywood panel?

6. When plywood is used as wall sheathing on a stud spacing of 16-in. OC, the thickness of the plywood must be at least how thick?

7. What is a simple type of horizontal joint in joining exterior panels?

8. In addition to its many other uses, the particle or chip board when used with high-pressure plastic laminates is primarily used as what?

9. Name the three types of hardboard.

10. By mechanical or chemical processes, wood logs are converted into what three things?

11. When using plywood for roof or wall sheathing, what controls the thickness?

12. Name two properties wood fiberboards offer.

ASSIGNMENT

1. Sketch a 5-ply plywood panel and label or name all the laminations.

2. Make a sketch showing a use of plywood other than those illustrated in this unit.

SUPPLEMENTARY INFORMATION

Laminated Fiberboard Decking

This decking is manufactured by laminating two or more fiberboards together. It is available in 5/8 to 3-in. thicknesses, 2'-0" and 3'-0" widths, and 4'-0" to 12'-0" lengths, with a tongue-and-groove finish on all four edges.

The thicker the decking, the greater the insulation value and the spanning between supports, and also the nail-holding power. When wood rafters are exposed, this material is the sheathing, insulating, and finish surface between the rafters (see Figure 19-7).

Insulating Board

Fiber insulating boards are manufactured in 1/2-in, 5/8-in., 3/4-in., and 1-in. thicknesses, 4'-0" wide,

and from 4'-0" to 16'-0" in length. This type of board is also used for manufacturing fiber-type acoustical tile, and is used in building construction for sheathing in residential buildings and as roof insulation for types of buildings with flat roofs or slightly pitched roofs.

Concrete Forms

Heavy paper in spiral form is manufactured in almost any diam-

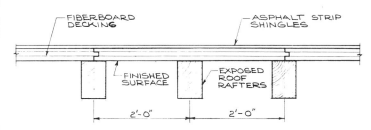

Figure 19-7. Laminated Fiberboard Decking with Exposed Roof Rafters

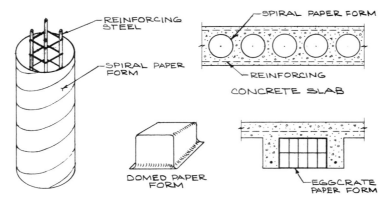

Figure 19-8. Paper Forms

eter for use as forms for reinforced concrete columns and slabs. It is also made into dome-type and egg crate-type forms (see Figure 19-8).

Wallpaper

Wallpaper is generally used for decorative or plain interior wall covering. It is manufactured in rolls and sheets of various widths and lengths, and is available with a wide variety of special finishes such as water-resistant, waterproof, and mark-resistant.

20

Laminated Wood and
Solid Wood Structural Members

INTRODUCTION

In this unit, the laminated wood structural members reviewed include beams, girders, timbers, decking, arches, trusses, and stressed-skin plywood panels. Solid wood structural members reviewed include the different types of trusses and timber connectors, as well as details of a nail-glued roof truss.

TECHNICAL INFORMATION

Laminated Wood Structural Members

Glue-laminated lumber is the product resulting from the joining together by glue of two or more smaller pieces of lumber to form a larger, stronger member. Generally, 1 or 2-in. thick stock is bonded to form beams, columns, arches, and various other structural members. The major advantage of such glued members lies in the fact that they are one-third stronger than a sawed member of equal size, as shown

graphically in Figure 20-1. This permits a slenderizing of design without sacrificing strength. Tests prove that the glued joint (the bond) is stronger than the wood itself, and time does not affect the bonding qualities of the glue. Another important advantage of laminated lumber is that larger pieces of lumber in both length and cross section may be obtained than is possible in single sawed pieces of solid wood. Furthermore, beams of varying cross section can be built to have a greater shear resistance at points where it is most desired.

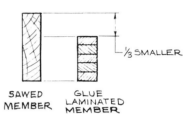

Figure 20-1. Sawed and Glue-Laminated Members of Equal Strength

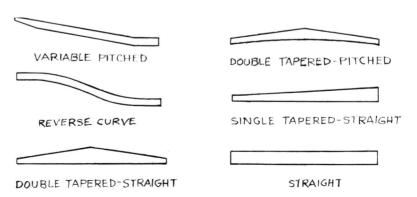

Figure 20-2. Glue-Laminated Beams

163

Laminated Beams and Girders

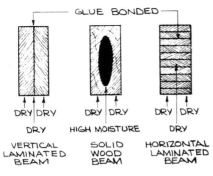

Figure 20-3. Uniform Moisture Control

Because laminated beams are manufactured in a plant where they are assembled, glued, planed, sawed, treated with preservatives, and finished with the natural wood grains showing, no plastering or painting or other decorative treatments are necessary unless they are specified to conform to the requirements of the architectural design. These beams are economically practical when used as a finished product and exposed as a part of the architectural design, but when used merely as concealed structural units, they are generally uneconomical. For spans up to 24 ft., solid timbers are usually cheaper than laminated sections except where extremely heavy loads are to be supported. Beyond that limit, glue-laminated beams or wood trussed are generally used. Figure 20-2 illustrates a wide variety in shapes of glue-laminated beams.

Laminated Timbers

Such timbers are superior to solid wood structural members. First, they have greater strength and rigidity than solid beams because they are vertically laminated to resist twisting, and second, there is uniform moisture control because the members are kiln-dried before laminating to a 12 to 15% moisture content (see Figure 20-3). Laminated timbers are manufactured in lengths from 12 to 60 ft. in increments of 1 ft. All timbers are given protective treatment with a water-repellent sealer.

Laminated Beam Support Connections

Laminated beams may be supported at their ends by masonry walls, laminated columns, simple frame wall construction, or steel columns. Figure 20-4 illustrates such supports together with the connection rough hardware for fastening the beams.

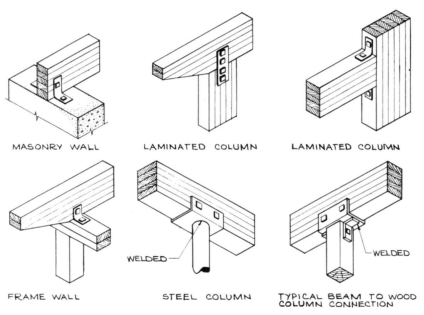

Figure 20-4. Laminated Beam Support Connections

Laminated Decking

The decking here is part of a structural concept that utilizes post and beam or arch construction with exposed roof decking (see Figure 20-5). This system contributes to simplified building through the use of fewer supporting elements than required in conventional frame construction. Laminated decking provides a V-grooved paneled ceiling with natural wood finish for spans between beams of up to $8'-0''$. Before the lumber used for decking is bonded, it is kiln-dried to

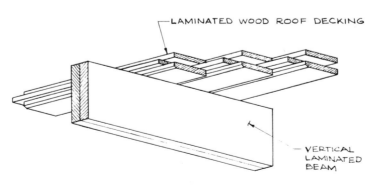

Figure 20-5. Laminated Beams Supporting Laminated Decking

a completely uniform moisture content, thus assuring dimensional stability. Decking is available in laminated thicknesses of 2-1/4 in., 2-3/4 in., 3 in., 3-3/4 in., and 4 in. (see Figure 20-6).

Two-hinged and Three-hinged Arches

Laminated two- and three-hinged arches are manufactured similarly to wood laminated beams and girders. They are curved and glued together under high pressure to make a solid and rigid member. Control of moisture content and the use of structural glues with a high water resistance insure a stable and permanently bonded structural member.

Laminated arches provide a means of roofing large areas without obstructions from floor to ceiling. Arch units may be spaced as much as 15 to 20 ft. apart, in which case purlins are used, which in turn support the roof deck material.

The two-hinged arches are particularly adaptable for wide spans. They have been used for spans up to 250 ft., and are practical for even larger spans. However, the most common spans range from 30 to 100 ft.

These structural units are installed by various methods, as shown in Figure 20-7, including foundation-anchored, tied, buttressed, or combinations of these methods. With foundation buttresses, the horizontal thrust is contained by the foundations. With tie rods below grade, to counteract the horizontal thrust, simple vertical foundations can be used. Tied arches are supported by columns or bearing walls, with headroom being obtained by having the tie rods located at the top of the columns or walls. Buttressed arches are contained by concrete or masonry buttresses of sufficient size to counteract the horizontal thrust of the arches.

Three-hinged arches (see Figure 20-8) are widely used in

churches, gymnasiums, auditoriums, and other types of buildings

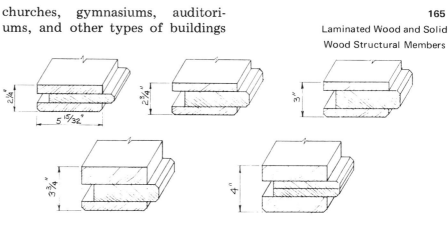

Figure 20-6. Laminated Decking

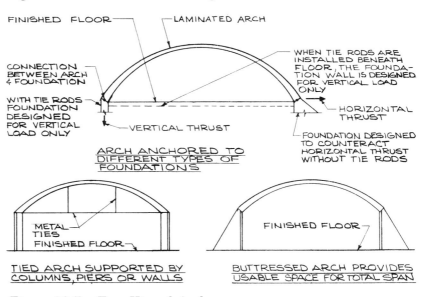

Figure 20-7. Two-Hinged Arches

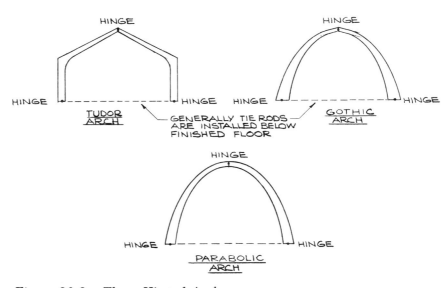

Figure 20-8. Three-Hinged Arches

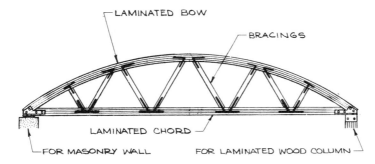

Figure 20-9.   A Typical Bow String Truss

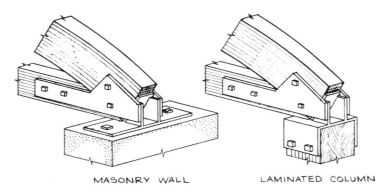

MASONRY WALL   LAMINATED COLUMN

TYPICAL HEEL CONNECTIONS

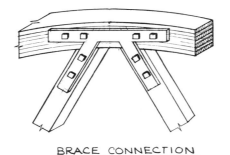

BRACE CONNECTION

Figure 20-10.   Laminated Wood Truss Connections and Supports

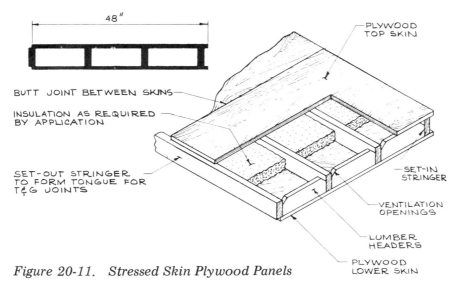

Figure 20-11.   Stressed Skin Plywood Panels

where large clear open areas are required. Specific types include the Tudor, the Gothic, and the parabolic. They are composed of two laminated units, fastened at the top or ridge of the arch and at its two foundation ends. Functionally, they provide clear span framing for both sidewalls and roof. Generally the horizontal thrust is counteracted by installing the rods below the finished floor.

Arches of this type are generally left exposed, because the natural wood form and grained surface can become a design feature of the interior or the exterior.

Laminated Wood Trusses

A wood truss is a complete structural unit with laminated wood top and bottom chords. Web members are solidly bolted between the laminated members with heavy steel bolts and straps. Trusses have an advantage over the arches in that they do not require ties, special foundations, or buttresses. Commonly used spans range from 40 to 150 ft., and some manufacturers are equipped to produce spans greater than 150 ft. Of the numerous types of wood trusses available, the bow-string type (see Figure 20-9) is perhaps the most common. Wood trusses may be supported by steel, concrete, or wood columns, or by masonry walls or piers. Typical examples of support and connection are shown in Figure 20-10.

Stressed Skin Plywood Panels

Plywood panels are nailed and glued to thin stringers to form an integral member capable of resisting stresses (see Figure 20-11). Such panels are used extensively in prefabricated building construction for floors and roofs because they permit thinner and shallower framing members than does conventional frame construction. All panels are standard 4-ft. wide plywood with the face grain parallel to the length. The lengths vary depending on their loading. Normally spans range up to 12 ft., but can be used for larger spans of 16 to 18 ft. When used as roof panels, the top and bottom plywood skins are 3/8 in. in thickness of exterior plywood. In panels used as floors, the top skin is 5/8 in. in thickness, while the lower skin is 3/8 in. in thickness; both skins are of interior plywood. The structural action of the stressed-skin panel is similar to the steel beam. The top flange of the beam when loaded tends to bend and set up a compressive stress while its bottom face is in tension. These actions tend to cause the plywood to slip from the beams; therefore, because nails alone cannot prevent such movement, the panels are glued to the tops and bottoms of beams under a pressure of 150 lbs. per square in. The type of glue used is very important — when panels are exposed to moisture, a highly moisture-resistant type of glue must be used.

Solid Wood Trusses

The type of built-up wood roof trusses used in a building is determined by the length of the required span, the material of which the truss is constructed, and the manner in which the truss is loaded. Several designs, all of which provide a method of roofing clear space, are shown in Figure 20-12. They are suitable for residential construction, housing, and industrial and commercial construction.

Typical lightweight trusses such as is used in residences are shown in Figure 20-13. These trusses can be prefabricated in a shop, as the larger ones usually are, or can be built on the site. In the latter case, they are fabricated on the ground, using lightweight material, usually 2 × 4s and 2 × 6s, generally placed 2'-0" on centers on the exterior frame walls. The parts of trusses

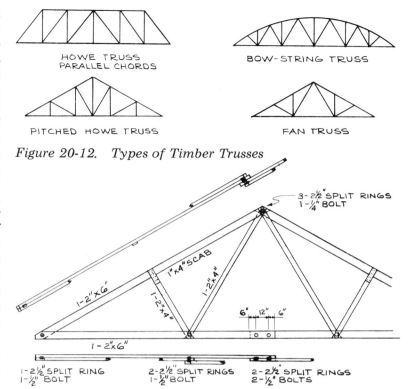

Figure 20-12. Types of Timber Trusses

Figure 20-13. Solid Wood Truss

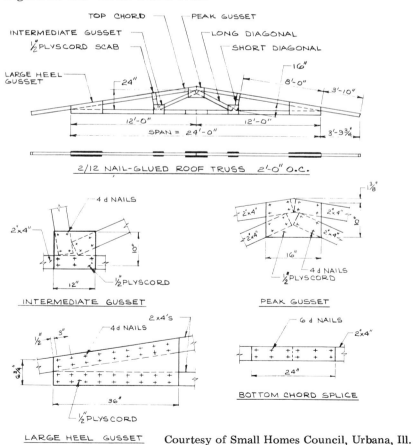

Courtesy of Small Homes Council, Urbana, Ill.

Figure 20-14. 2/12 Pitch Nail-Glued Roof Truss

167

can be joined with either plywood connectors or suitable metal timber connectors at critical points. The roof sheathing is nailed directly to the trusses. Figure 20-13 shows a solid wood truss joined with split-ring connectors. Such trusses are used for spans from 20 to 35 ft. in length. Figure 20-14 shows the construction details of a nail-glued roof truss.

REVIEW EXAMINATION

1. What is the advantage of a glue-laminated beam over a sawed member?

2. What is the thickness of the boards used in fabricating a laminated beam?

3. Laminated timbers are manufactured in lengths ranging from 12 ft. to 60 ft. in what increments?

4. Laminated decking is available in what thicknesses?

5. Practical and most common spans of laminated two-hinged arches range from what span to what span in ft.?

6. What are two-hinged arches particularly adaptable for?

7. What is the advantage of laminated wood trusses over arches?

8. What is the structural action of the stressed-skin panel similar to?

9. What is the purpose of trusses?

10. Name the roof truss considered to be one of the most efficient for spans exceeding 80 ft.?

ASSIGNMENT

1. Sketch three laminated beam support connections.

2. Sketch a two-hinged and a three-hinged laminated arch, showing methods of counteracting horizontal thrust.

SUPPLEMENTARY INFORMATION

Design Considerations

Glue-laminated structural members are available in widths of 3-1/4, 5-1/4, 7, 9, 11, and 12-1/2 in. and in beam depths ranging from 3-1/8 to 92-1/2 in.

Following is an example of how sectional dimensions are calculated, together with the section properties of a 5-1/4-in.-wide glue-laminated member of various beam depths and number of laminations (see Table 20-1).

Timber Connectors

When bolts are used in solid timber framing, the stresses are concentrated on the relatively small areas on which the bolts bear. Use of timber connectors affords a comparatively large area of wood against which the connector exerts pressure, and thus the stresses are distributed over almost the entire cross-sectional area of the timbers that are connected. The connectors transmit loads from one member to another with a minimum reduction of the cross-sectional area of the joined members, and they permit the use of small pieces of lumber in the construction of solid wood trusses.

The various types of connectors in common use are the split ring, the toothed ring, the claw plate, and the shear plate (see Figure 20-15).

Nail-glued Headers over Openings

These consist of two upper flanges, 2 × 4-in., and a single 2 × 4-in. lower flange separated by vertical 2 × 4-in. spacers 2'-0" on centers (see Figure 20-16). A skin of plywood is glued and nailed to the framing members on one or both sides of the framing members. The nail-glued header serves as a lintel over larger openings which is capable of supporting roof loads greater than lintels of solid framing lumber.

Figure 20-17 illustrates the header spans attainable under various design loads for type A and B headers when the overall truss lengths are known.

The Rigid Plywood Frame

Figure 20-18 is composed of two vertical members and two rafters of 2 × 8-in. nominal size lumber. The rafters have a 5 to 12 pitch and are fastened at their crown by plywood gusset plates on two sides (see Detail A) and to the frame leg at haunch (see Detail B). The frame is attached to a sill plate, which is the same size as the vertical or leg member (see Detail C).

The number of nails used at the connections, the size of the nails, and their spacing is important, because this patented frame is engineered to precise specifications. For center-to-center distances between frames, consult the Douglas Fir Plywood Association, Tacoma, Washington.

Table 20-1. *How to Calculate Sectional Dimensions*

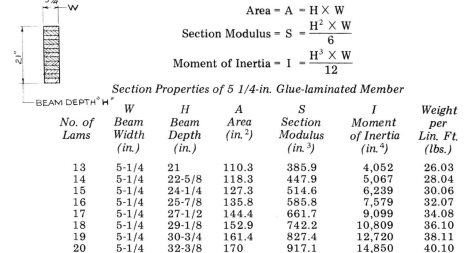

$$\text{Area} = A = H \times W$$
$$\text{Section Modulus} = S = \frac{H^2 \times W}{6}$$
$$\text{Moment of Inertia} = I = \frac{H^3 \times W}{12}$$

Section Properties of 5 1/4-in. Glue-laminated Member

No. of Lams	W Beam Width (in.)	H Beam Depth (in.)	A Area (in.²)	S Section Modulus (in.³)	I Moment of Inertia (in.⁴)	Weight per Lin. Ft. (lbs.)
13	5-1/4	21	110.3	385.9	4,052	26.03
14	5-1/4	22-5/8	118.3	447.9	5,067	28.04
15	5-1/4	24-1/4	127.3	514.6	6,239	30.06
16	5-1/4	25-7/8	135.8	585.8	7,579	32.07
17	5-1/4	27-1/2	144.4	661.7	9,099	34.08
18	5-1/4	29-1/8	152.9	742.2	10,809	36.10
19	5-1/4	30-3/4	161.4	827.4	12,720	38.11
20	5-1/4	32-3/8	170	917.1	14,850	40.10

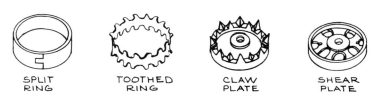

SPLIT RING   TOOTHED RING   CLAW PLATE   SHEAR PLATE

Figure 20-15. *Timber Connectors*

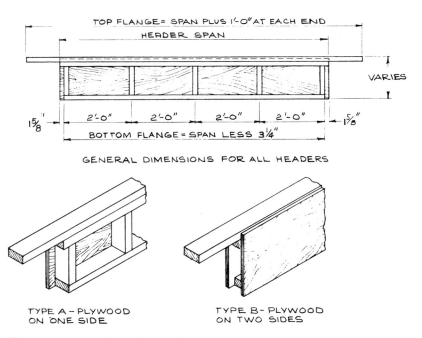

TOP FLANGE = SPAN PLUS 1'-0" AT EACH END

HEADER SPAN

2'-0"   2'-0"   2'-0"   2'-0"

BOTTOM FLANGE = SPAN LESS 3¼"

VARIES

1⅝"

1⅝"

GENERAL DIMENSIONS FOR ALL HEADERS

TYPE A – PLYWOOD ON ONE SIDE

TYPE B – PLYWOOD ON TWO SIDES

Figure 20-16. *Nail-Glued Header*

This type of roof construction is adaptable for wide spans and economy of construction. Horizontal thrust is contained by a tension ring at the periphery of the structure so that all members are in compression. Glulam (glued laminated) segments of the dome are of small, economical section dimensions, and the walls need not be buttressed. Ratio of rise to span is low, minimizing the amount of decking required. Dome patterns are radial rib, triangular, and triax; the latter two are assembled on the ground, then raised into position and connected to the tension ring, thus saving the cost of scaffolding. Dome patterns in plan are shown in Figure 20-19.

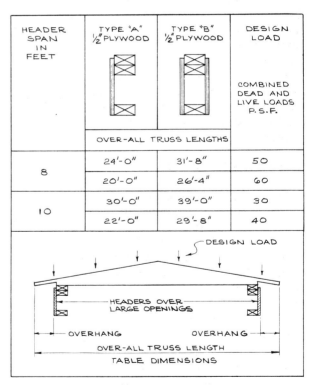

HEADER SPAN IN FEET	TYPE "A" ½" PLYWOOD	TYPE "B" ½" PLYWOOD	DESIGN LOAD
			COMBINED DEAD AND LIVE LOADS P.S.F.
	OVER-ALL TRUSS LENGTHS		
8	24'-0"	31'-8"	50
	20'-0"	26'-4"	60
10	30'-0"	39'-0"	30
	22'-0"	29'-8"	40

Figure 20-17.   Header Spans

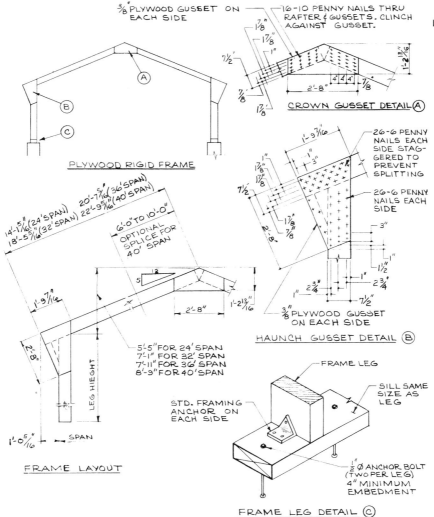

PLYWOOD RIGID FRAME

CROWN GUSSET DETAIL (A)

HAUNCH GUSSET DETAIL (B)

FRAME LAYOUT

FRAME LEG DETAIL (C)

Courtesy of Douglas Fir Plywood Assoc., Tacoma, Washington

Figure 20-18. Plywood Rigid Frame

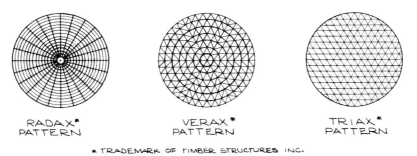

RADAX*
PATTERN

VERAX*
PATTERN

TRIAX*
PATTERN

* TRADEMARK OF TIMBER STRUCTURES INC.

Figure 20-19. Laminated Wood Domes

21

Building with Wood

INTRODUCTION

This unit deals with some of the major details used in contemporary wood frame construction based on laminated wood bents and a unique system in which two milled members are employed to form the major part of the construction. Also included are light frame conventional sections and details, and post and beam construction.

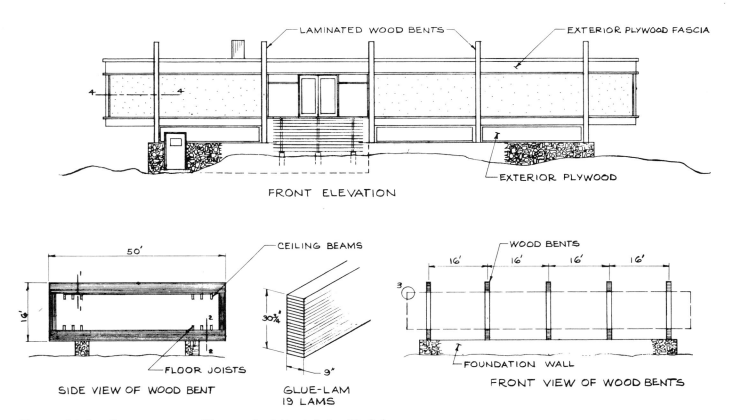

FRONT ELEVATION

SIDE VIEW OF WOOD BENT

GLUE-LAM
19 LAMS

FRONT VIEW OF WOOD BENTS

Figure 21-1.   Contemporary House, Architect John M. Johansen

173

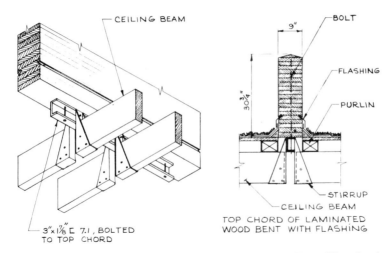

Figure 21-2. *Ceiling Beams Supported from Top Chord of Laminated Bents, Section 1-1 (see Fig. 21-1)*

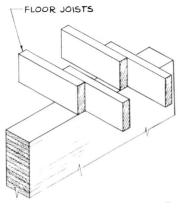

Figure 21-3. *Floor Joists Resting on Bottom Chord of Laminated Bents, Section 2-2 (see Figure. 21-1)*

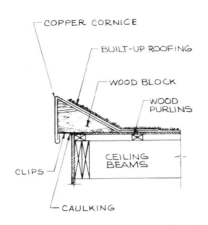

Figure 21-4. *Cornice at Roof, Detail 3 (see Fig. 21-1)*

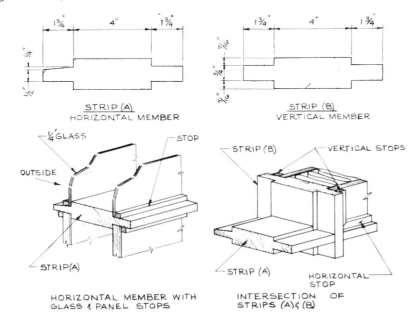

Figure 21-5. *Milled Section*

TECHNICAL INFORMATION

The Contemporary House

A contemporary house (see Figure 21-1) employs a system of framing and construction consisting of five laminated wood bents in hollow rectangular shape from which the roof plane hangs and upon which the floor structure rests.

Ceiling beams. The ceiling beams are suspended from the laminated bents by means of joist hangers which are attached to a channel iron, bolted to the top cord of the laminated wood bents (see Figure 21-2, Section 1-1). The top and bottom cords of the wood bents are composed of nineteen wood members, glued and laminated together under high pressure, forming a $9 \times 30\text{-}3/4$-in. beam. The floor joists are resting on and span the lower cords of the wood bents (see Figure 21-3, Section 2-2).

Roof construction. Roof purlins running at right angles to the ceiling beams provide the nailing for the roof sheathing and roofing material. The roof cornice (see Figure 21-4, Detail 3) is of a beveled wood blocking finished with copper fascia. The roof is a built-up roofing.

Wall construction. Instead of the usual stud wall, there are a number of milled sections (see Figure 21-5) which, when in place, will receive all doors, windows, glass, and other wall panels. When the floor or platform is in place, "strip A" is placed around the outside edges of the floor instead of the usual shoe or sole plate. This immediately provides the exterior bottom trim with a drip or shadow line under the striated plywood panels used on the outside faces of the walls.

In Figures 21-6 and 21-7 show sections and isometric drawings illustrating how "strip A" can take the place of the usual 2×4 sole plate on both frame and masonry

174

sills. "Strip A," because it projects above the finish floor when it is laid in place, acts as the baseboard. The walls can be insulated as usual, and the exterior and interior panels and finish can be applied.

Vertical 4 × 4s are capped by another "strip A," nailed directly on top of the posts. This strip also can act as a head trim for all doors and windows. All vertical members, "strip B," serve as vertical mullions and post casings, because the building on which these milled sections are used are ideally suited to post and beam construction. Around all door openings, double vertical "strip Bs" are used for additional stability. Filler pieces are provided for closing in where two "B" strips are used (see Figure 21-8).

The wall section (see Figure 21-9) illustrates the use of "strip A" and "strip B" from foundation to roof.

Figure 21-10 illustrates the extension of the house beyond the laminated wood bents (see Figure 21-1, Section 4-4). The protruding area serves as a porch with roof overhead and two side walls. Note the "B strips," the vertical members providing for exterior and interior finish. Panels of glass are fixed at the two side walls, and there are sliding glass doors in the middle leading to the cantilevered porch.

Conventional Wood Frame Construction

Frame structures for houses may be constructed by four different methods known as balloon framing, braced framing, western or platform framing, and the modern braced framing. Although these methods are essentially the same, the western or platform framing is the most commonly used of the four methods (see Figure 21-11). In this method, after placing of the sill header and floor joists, the subflooring is installed, forming the platform. On this platform, all exterior and interior bearing walls are

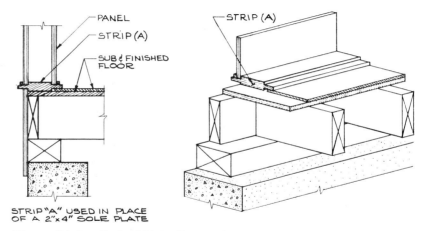

Figure 21-6. Strip (A) in Frame Construction

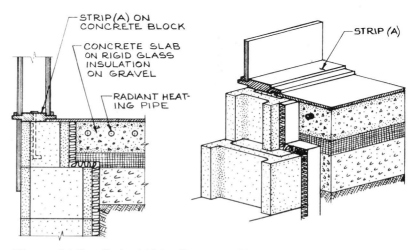

Figure 21-7. Strip (A) in Concrete Slab on Grade Construction

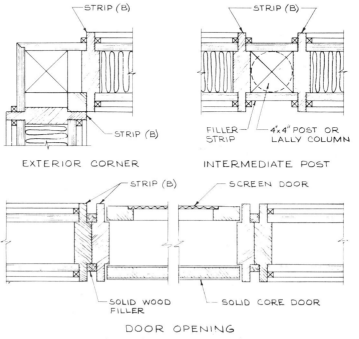

Figure 21-8. Details Illustrating the Use of Strip (B)

175

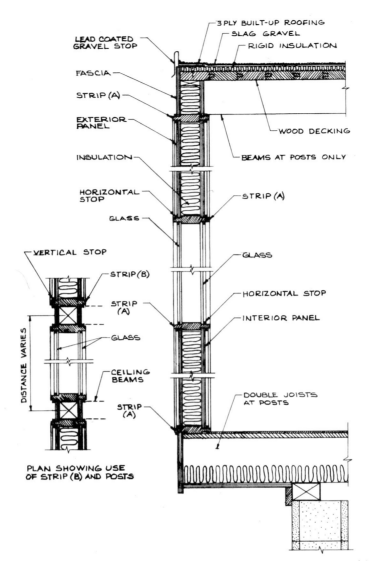

Figure 21-9. Post and Beam Construction Using Strips (A) and (B)

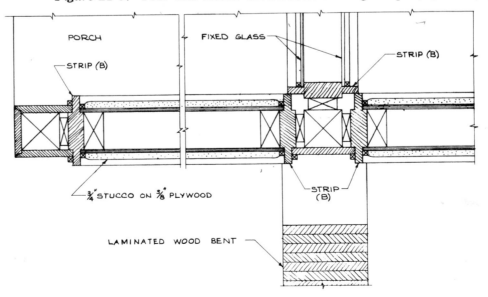

Figure 21-10. Section 4-4 (see Fig. 21-1)

installed, using 2 × 4-in. studs spaced usually 16-in. OC, and set on a sole plate and capped with a girt of two 2 × 4-in. members. The second floor joists are next placed in the girts of the bearing walls. The subflooring forms the second-floor platform. In like manner, all exterior and interior bearing walls are installed to receive the ceiling beams and roof rafters.

Light wood framing details are shown in Figures 21-12 and 21-13.

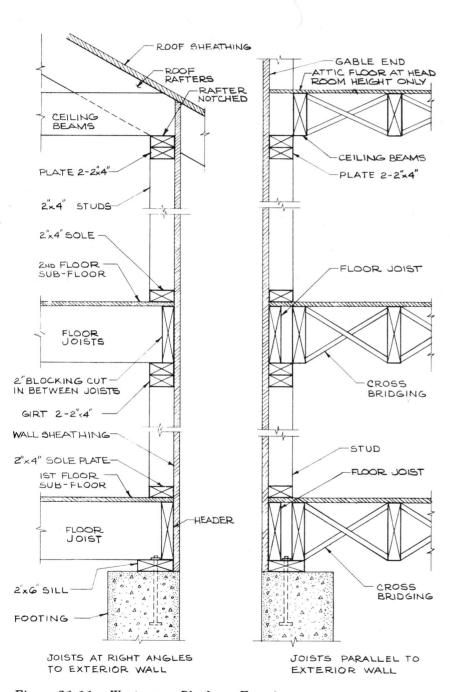

Figure 21-11.   Western or Platform Framing

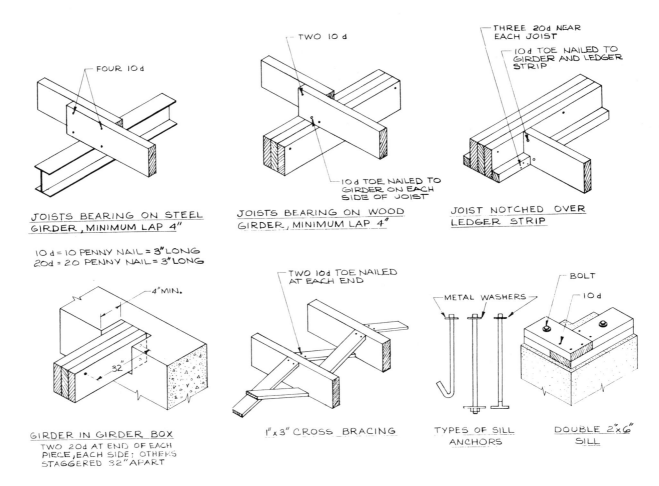

JOISTS BEARING ON STEEL
GIRDER, MINIMUM LAP 4"

10 d = 10 PENNY NAIL = 3" LONG
20 d = 20 PENNY NAIL = 3" LONG

JOISTS BEARING ON WOOD
GIRDER, MINIMUM LAP 4"

JOIST NOTCHED OVER
LEDGER STRIP

GIRDER IN GIRDER BOX
TWO 20d AT END OF EACH
PIECE, EACH SIDE; OTHERS
STAGGERED 32" APART

1"x 3" CROSS BRACING

TYPES OF SILL
ANCHORS

DOUBLE 2"x6"
SILL

Figure 21-12. Light Wood Framing Details

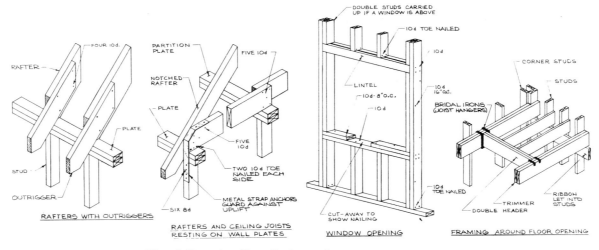

RAFTERS WITH OUTRIGGERS

RAFTERS AND CEILING JOISTS
RESTING ON WALL PLATES

WINDOW OPENING

FRAMING AROUND FLOOR OPENING

Figure 21-13. Light Wood Framing Details (cont.)

1. When two wood floor joists are supported on a wood girder, what is the minimum dimension for bearing?

2. When framing around openings, what is used to support the header beam?

3. In the contemporary house the entire construction system is supported by what?

4. The advantage of a milled section as illustrated is that it can serve multiple purposes. Name three.

5. Frame structures for houses may be constructed in what four different methods?

6. What is the most common and often-used of the above methods?

ASSIGNMENT

1. Sketch a detail showing how the top bent of the contemporary house supports the ceiling beams.

2. Sketch the western or platform frame wall section.

SUPPLEMENTARY INFORMATION

Post and Beam Construction

In post and beam construction (see Figure 21-14), the roof beams may be placed transversely where the top end of the beam is fastened against a ridge and the lower end of the beam is resting on a post 4 × 4-in. minimum size. The rafters may be spaced from 2'-0" to 4'-0" on centers depending on their spans. Roof planks which span the rafters serve as the roof sheathing and, in many cases, as the inside finished ceiling. Post and beam construction may also be accomplished when the roof beams are placed longitudinally and supported intermittently or at their ends by posts.

Advantages of post and beam construction are that it provides a greater flexibility in interior room planning and allows the use of large panel wall coverings between the structural post supports.

Nails

Table 21-1 gives the size, diameter, length, and number of nails per lb. for wood frame construction.

Table 21-1. Nails

Size penny	Diameter (in inches)	Length (in inches)	No. of Nails per Pound
2d	11/64	1	847
3d	13/64	1-1/4	543
4d	1/4	1-1/2	296
5d	1/4	1-3/4	254
6d	17/64	2	167
7d	17/64	2-1/4	150
8d	9/32	2-1/2	101
9d	9/32	2-3/4	92.1
10d	5/16	3	66
12d	5/16	3-1/2	56.1
16d	11/32	3-1/2	47.4
20d	13/32	4	29.7
30d	7/16	4-1/2	22.7
40d	15/32	5	17.3
50d	1/2	5-1/2	13.5
60d	17/32	6	10.7

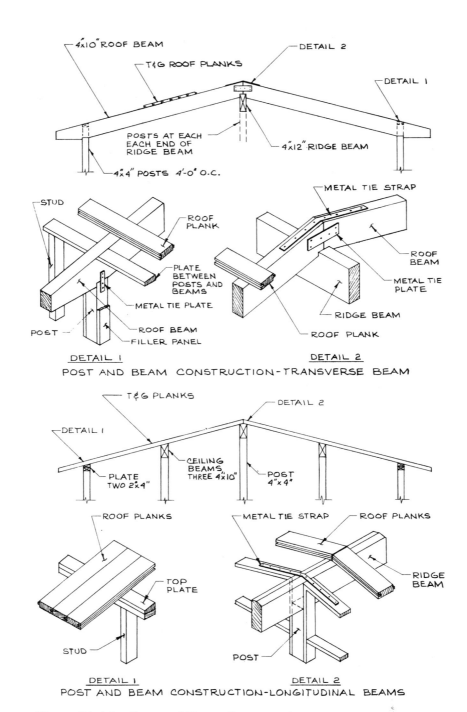

Figure 21-14.   Post and Beam Construction

22

Metals

INTRODUCTION

This unit deals with the metals that are extensively used in construction: iron, steel, stainless steel, aluminum, brass, bronze, copper, and zinc. These metals are divided into two major groups, ferrous and nonferrous.

TECHNICAL INFORMATION

Characteristics of Metals

Ferrous, from the Latin word for iron, includes iron and the iron-based metals such as various steels and stainless steel.

Aluminum, brass, bronze, copper, and zinc are nonferrous metals. There are others in this group such as tin, which at one time was widely used for roofing and siding; gold and silver, which have decorative uses; and chromium, titanium, and others, which are ingredients in alloys. Here the emphasis is on metal used in construction.

As a group, metals are substances that have a hardness, can conduct heat and electricity, and possess certain mechanical properties. The most important characteristic is their power to resist change in

shape. Almost all metals used in construction are alloys and not the pure metals. Pure iron, for example, is as soft and bendable as copper.

Most metals deteriorate as a result of the slow action of substances in the air, soil and water, or of chemical agents to which they are exposed. This is called *corrosion*. The rusting of iron and steel are examples of corrosion. Another type of deterioration is galvanic action.

Galvanic action. When different metals or alloys are joined or brought into contact with each other and moisture is present, a galvanic action starts — an electrical current starts to flow from one metal to the other and in time one of the metals is eaten away while the other remains intact. How long this takes depends on the amount of moisture present, even if it is only the humidity in the air. In or near the ocean or in the presence of water containing minerals, galvanic action can become much more intense.

For example, if aluminum siding is applied with steel nails, the aluminum around the steel will gradually be eaten away and the siding will eventually fall off. In order to

Table 22-1. Galvanic Table

Aluminum
Zinc
Iron and Steel
Stainless steel
Lead
Brasses
Copper
Bronze
Gold

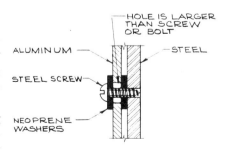

ISOLATION WHERE METALS CAN AND WILL BE ATTACKED BY GALVANIC ACTION

overcome this galvanic action, different metals must be isolated or applied with metals that are compatible and will not start this electrical action. The galvanic table (Table 22-1) shows the metals listed in a sequence in which each metal is attacked through galvanic action by all the metals that follow it in the table. Thus, aluminum is attacked by all metals listed, whereas gold is not attacked by any of the metals. Steel nails cannot be used to secure stainless steel; the stainless steel will, through galvanic action, disintegrate the steel nails.

Uses of Metal in Buildings

Metals are used in six major categories in the construction of buildings, as outlined in Table 22-2.

Weights of Metals

Metals vary considerably in their weight, as can be seen in Table 22-3. In the construction field, however, the weight of the metals or building materials becomes important only in the multistoried building. Consideration of relative weights then becomes the problem of the structural engineer and also must be considered by the professional estimator, who calculates the quantities of materials required from the working drawings for a building.

An article that weighs 6 lbs. in aluminum will weigh 6 times 3.17, or 19.02 lbs. in brass. The weight of an article in the other metal must be divided by the conversion factor to find out what it would weigh in aluminum — a copper article weighing 25 lbs. would weigh 25 divided by 3.28, or 7.62 lbs. in aluminum.

Ferrous Metals

Iron, steel, and stainless steel. The element *iron* is used to make wrought and cast iron, steel, and stainless steel. The characteristics of these irons and steels depend upon the iron-carbon percentages and the addition of small quantities of other elements such as nickel, chromium, manganese, boron, aluminum, copper, and silicon (see Tables 22-4 and 22-5).

Table 22-2. Uses of Metal in Buildings

Category	Metal	Where Used
1. Structure	Steel	Structural steel, reinforcing of concrete
2. Hollow metal work (including curtain walls)	Steel, stainless steel, aluminum, brass, and bronze	Doors, bucks, partitions, panels, windows, mullions, curtain walls
3. Miscellaneous metal work	Steel, stainless steel, aluminum, wrought iron, cast iron, brass, and bronze	Railings, fencing, stairs, lintels, gratings, ladders, hardware
4. Ornamental metal work	Stainless steel, aluminum, brass, and bronze	Plaques, letters, grilles, railings, screens, sculpture, hardware
5. Flashing	Stainless steel, zinc, copper, aluminum	Cap and base flashing, spandrel flashing, copings, caps, leaders and gutters
6. Miscellaneous	Steel, stainless steel, aluminum, copper, brass, bronze, zinc	Nuts, bolts, rivets, screws, nails, wire, cable

Table 22-3. Weights of Metals

Metal	Specific Gravity	Weight Pounds per 10 Cubic Inches	Weight Conversion Factor
Iron, steel, and stainless steel	7.80 to 7.90	2.82 to 2.85	2.9
Aluminum	2.70	0.975	1.0
Copper	8.89	3.210	3.28
Brass	8.20 to 8.60	2.961 to 3.105	3.17
Bronze	8.78	3.171	3.27
Zinc	7.04 to 7.16	2.54 to 2.59	2.64

Table 22-4. Percentages of Carbon

Wrought Iron	Cast Iron	Carbon Steel	Stainless Steel
Less than 0.1% carbon	More than 2.0% carbon	Less than 2.0% carbon	Less than 0.2% carbon

Table 22-5. Percentage of Generally Used Alloying Elements Used in Steel and Stainless Steel

Alloying Element	Carbon Steel	Stainless Steel
Nickel	———	6.0% to 14.0%
Chromium	with aluminum and boron less than 3.99%	14.0% to 19.0%
Manganese	less than 1.65%	2% maximum
Copper	less than 0.60%	———
Silicon	less than 0.60%	———

Cast and wrought iron. *Cast iron* is a hard and brittle material which can be cast into almost any shape and can be welded by any of the generally used methods. In the field of construction it is used for dampers, manhole covers, gratings, circular stairs, pipe, plumbing fixtures, hardware, etc.

Wrought iron is a soft, tough material which is resistant to corrosion. It can be bent, rolled, machined, and spun, and it can be welded by any of the generally used methods. In the field of construction it is used for railings, ornamental ironwork, grilles, pipe, and furniture. It is manufactured in pipe, rod, bar, plate, sheet, and bent shapes.

Steel. Steel is a hard, strong material which can be rolled, drawn, bent, and cast; it can be welded by any of the generally used methods. In the construction field it is extensively used for structural framing, as reinforcing and forms for concrete, and in hollow metal work, miscellaneous metal work, and ornamental metal work, etc.

Structural framing consists of various rolled sections, bent shapes, tubular shapes and open-web joists as shown in Figures 22-1, 22-2, 22-3, and 22-4.

Reinforcing and forms for concrete. Steel reinforcing bars and wire mesh are used for reinforced concrete and steel wire rope is used for prestressed concrete (see Tables 22-6 and 22-7 and Figures 22-5 and 22-6).

There are many types of forms for concrete and reinforced concrete, such as forms for foundation walls, steel decking, corrugated steel, steel pans, and domes (see Figure 22-6).

Stainless steel. This is a very hard, stiff, strong, and corrosion-resistant material that can be machined, cast, rolled, drawn, bent, formed and welded, and soldered. Stainless steel is manufactured in sheet, strip, plate, bar, tubing, castings, and

structural shapes. It is widely used in the construction field for doors, windows, curtain walls, signs, grilles, louvres, railings, flashing, and structures in which it is exposed to the atmosphere. There is a wide variety of finishes from a mirrorlike finish to a rough hammered finish.

Nonferrous Metals

Aluminum. All aluminums used in building construction are alloys and vary from soft to almost as strong as steel; all are very corrosion-resistant. Aluminum is a lightweight metal and therefore has many advantages over other metals such as copper, brass, and steel. Its comparative lighter weight reduces shipping, handling, and installation costs (see Table 22-3).

Aluminum alloys are divided into two main categories: those for casting and those for wrought aluminum products. Wrought aluminum can be extruded, bent, formed, rolled, spun, machined, and welded and soldered by special methods. It is manufactured in sheet, strip, plate, rod, bar, pipe, tube, wire, foil, and numerous extruded shapes both standard and specially designed (see Figure 22-7). Aluminum is used for windows, doors, curtain walls, railings, screens, louvres, signs, plaques, copings, and flashings.

Corrosion of aluminum. Aluminum in contact with other metals

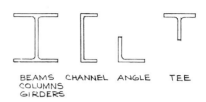

Figure 22-1. Rolled Sections

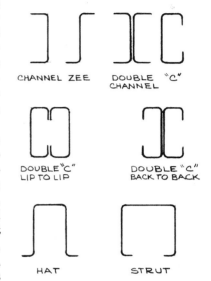

Figure 22-2. Bent Shapes

Figure 22-3. Tube Shapes

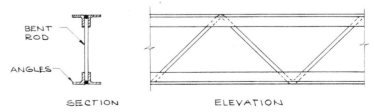

Figure 22-4. Open Web Steel Joist

that create galvanic action or with materials that create a chemical action can corrode when a moisture condition is present. To offset this possibility, proper protective measures must be taken. The aluminum should be isolated from these metals and materials by being

Table 22-6. Plain and Deformed Reinforcing Bars

Number	Diameter		Number	Diameter	
2	0.250	1/4"	7	0.875	7/8"
3	0.375	3/8"	8	1.000	1"
4	0.500	1/2"	9	1.128	1-1/8"
5	0.625	5/8"	10	1.270	1-1/4"
6	0.750	3/4"	11	1.410	1-3/8"

Table 22-7. Reinforcing Wire Mesh

Commonly Used Wire Mesh for Reinforcing Concrete

Size of Mesh	Gauge of Wire	Weight in Pounds for 100 sq. ft. of Mesh 5'-0" Wide
6" × 6"	No. 10	21
6" × 6"	No. 8	30
6" × 6"	No. 6	42

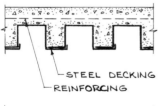

6 x 7 MEANS
6 STRANDS OF WIRE, EACH
STRAND CONTAINING 7 WIRES
STANDARD WIRE ROPE:
6x7, 6x9, 6x37, 8x19

Figure 22-5. Wire Rope

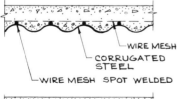

STEEL DECKING
REINFORCING

WIRE MESH
CORRUGATED STEEL
WIRE MESH SPOT WELDED

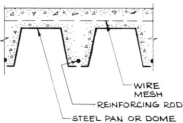

WIRE MESH
REINFORCING ROD
STEEL PAN OR DOME

Figure 22-6. Concrete Forms

coated with asphalt or by plastic (neoprene) washers or by a combination of both. When aluminum is used near the seacoast or in heavy industrial sections of the country, a noticeable amount of oxidation takes place. The rate of oxidation decreases substantially after about two years of weathering and then proceeds at a very slow rate. The penetration depth of this oxidation does not occur evenly all over the surface but develops in isolated spots (pitting). It spreads to new areas slowly, yet is not much deeper after twenty years than after one or two years.

Copper. Copper is a soft, corrosion-resistant, reddish metal used extensively in the building construction field for flashing, roofing, screens, and tubing. Because it is one of the best conductors of electricity, it is also used extensively in the electrical field. When exposed to the elements, copper develops a bright green coating which will then stop any further corrosion. Unfortunately, this coating can permanently stain adjoining materials, and care should always be taken to prevent such staining when copper is used.

Brass. When copper and zinc are combined, the result is an alloy called brass. Brass can vary in color, strength, and corrosion resistance depending on the percentage of copper and zinc (see Table 22-8).

Table 22-8. Color Variations of Brasses

Color	Copper	Zinc
Bronze color	90%	10%
Golden color	85%	15%
Yellow color	80%-62%	20%-38%
Silvery white	55%	45%

Brasses with 36 to 45% zinc are stronger but less corrosion-resistant, whereas the brasses with from 5 to 36% zinc are softer and more corrosion-resistant. Lead is added in small quantities to brasses to make them more workable.

Bronze. When copper and tin are combined, the resulting alloy is bronze. As originally developed, bronze was an alloy of 90% copper and 10% tin. This bronze is golden-brown in color, corrosion-resistant, strong and hard, but difficult to work. It was used primarily for casting. Now, most "bronzes" are alloys of copper, zinc, and small percentages of tin so that, in effect, there is little difference between "bronze" and "brass," and in today's terminology the words have almost become interchangeable if not synonymous. Brasses with the color of bronze are now known as architectural bronze, commercial bronze, and naval bronze. With these types of bronze, care should be taken in checking their specific characteristics in relation to where they are used in buildings.

Zinc. This is a medium-hard, low-strength, corrosion-resistant metal used in building construction for flashings, copings, roofing, hardware, and bathroom accessories. It also finds great use as a protective coating for iron and steel, called "galvanizing." When zinc is alloyed with copper it produces brass, and when alloyed with copper and nickel, it produces nickel silvers.

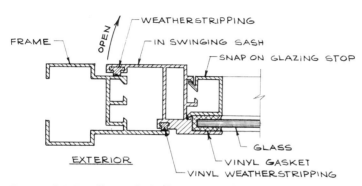

Figure 22-7. Extruded Window Sash and Frame Jamb Section

REVIEW EXAMINATION

1. What are the two major groups of metals?

2. Do metals resist change in shape?

3. When metals are exposed to the air, soil, water, or chemical agents, what happens?

4. When two different metals are in contact and moisture is present, what happens?

5. Cast iron, wrought iron, steel, and stainless steel depend upon what percentages of what two things for their characteristics?

6. What are the most important characteristics of stainless steel?

7. Name three areas where steel is used in the construction field.

8. What similar characteristics do aluminum and stainless steel share?

9. What is copper used for in the construction field?

10. Bronze is an alloy of copper and what?

11. Brass is an alloy of copper and what?

12. Name two areas in the construction field where zinc is used.

ASSIGNMENT

1. Make a sketch at 1/2-in = 1'-0" of several steel-rolled sections and tubular shapes.

2. In fifty words or less, describe where the different metals are used in the construction of buildings.

SUPPLEMENTARY INFORMATION

Carbon Steels

Carbon steels not only vary greatly by the percentages of iron and carbon but are also controlled by heat treatment and mechanical work during manufacture. The strength varies depending on whether it is slowly or rapidly cooled and whether it is hot- or cold-rolled. Thus carbon steels may be specified either by chemical composition, by method of manufacture and mechanical properties, or by a system of numerical designations. Table 22-9 lists the carbon

Table 22-9. Carbon Steels Used in Building Construction

Use	Carbon Percentages
Steel castings	limited to 0.12 to 0.50
Reinforcing for concrete	limited to 0.40 to 0.70
Structural shapes and small structural shapes	0.22
Cold-rolled steel (sheet)	not over 0.35
Wire for mesh and wire cloth	low
Wire for prestressed concrete	limited to 0.90 to 1.10
Rivets	limited to 0.05 to 0.20
Screws, nails, etc.	limited to 0.08 to 0.16

steels according to use in the construction field and in relation to carbon percentages.

Cor-ten Steel

Cor-ten steel is a high-strength, low-alloy steel which, when exposed to the atmosphere, develops a highly protective coating of a dark brown to a warm purple-black color. There are two types: Cor-ten A, for general architectural applications, and Cor-ten B, used where 50,000 PSI minimum yield point is required and always in thicknesses over 1/2-in.

These steels are used exposed for roofing, siding, windows, trim, structures, and ornamental designs. The minimum thickness for exposed Cor-ten steels is 18-gauge (0.0478 in.).

When these steels are incorporated in an architectural design, care should be taken that the adjacent materials used are of such a type and character that resist staining, because as the colored protective coating develops on the Cor-ten, some of it washes off and can cause staining.

Designations for Metal Products

Steel wire mesh reinforcing for concrete is indicated in this manner: 6.6–6.6 or 8.0–12.4. These symbols are used both on structural and architectural drawings as shown in Figure 22-8. To explain these symbols, let us take 6.6–6.6 mesh. The first 6 indicates that lengthwise wires are spaced 6 in. on center, and the second 6 indicates the size of lengthwise wires — in other words, that the lengthwise wires are No. 6 gauge. The third 6 indicates that the crosswires are spaced 6 in. on center, and the last 6 indicates that the crosswires are No. 6 gauge (see Figure 22-9). Similarly 8.0–12.4 mesh indicates that lengthwise wires are 8-in. OC of No. 0 gauge and crosswise wires are 12-in. OC of No. 4 gauge (see Figure 22-10).

Wire is designated by gauge number for ferrous and nonferrous types; it is also designated by diameter in decimals of an inch, as shown in Table 22-10. Note that the smaller the gauge number, the larger the diameter.

Sheet and strip metals have different methods of designation depending on the metal. For example, steel and zinc are designated by gauges, copper by weight per sq. ft. in ounces, and aluminum by alloy (see Tables 22-11, 22-12, 22-13, and 22-14).

Figure 22-8. Steel Wire Mesh Reinforcing

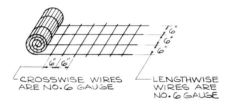

Figure 22-9. 6.6—6.6 Wire Mesh

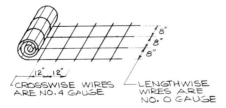

Figure 22-10. 8.0—12.4 Wire Mesh

Table 22-10. Wire Gauges of Ferrous and Nonferrous Wire

Gauge No.	Ferrous American Steel Wire Gauge (decimals of an inch)	Nonferrous American Wire Gauge (decimals of an inch)
00	0.3310	0.3648
0	0.3065	0.3249
1	0.2830	0.2893
2	0.2625	0.2576
3	0.2437	0.2294
4	0.2253	0.2043
5	0.2070	0.1819
6	0.1920	0.1620
7	0.1770	0.1443
8	0.1620	0.1285
9	0.1483	0.1144
10	0.1350	0.1019
11	0.1205	0.0907
12	0.1055	0.0808
13	0.0915	0.0720
14	0.0800	0.0641
15	0.0720	0.0571
16	0.0625	0.0508
17	0.0540	0.0453
18	0.0475	0.0403
19	0.0410	0.0359
20	0.0348	0.0320

Table 22-11. Zinc Designations

Gauge	Thickness (in inches)
9	0.018
10	0.020
11	0.024
12	0.028

Table 22-12. Copper Designations

Weight per Sq. Ft. (ounces)	Thickness (in inches)
16	0.0190
20	0.0245
24	0.0300
32	0.0405

Table 22-13. Aluminum Designations

Alloy	Thickness Available (inches)
1100–0	0.02, 0.025, 0.032
1100–H14	0.04, 0.051, 0.064
3003–H14	0.02, 0.025, 0.032

Table 22-14. Steel and Stainless Steel Designations

Gauge	Steel Thickness (in inches)	Stainless Steel Thickness (in inches)
8	0.1644	0.171875
9	0.1495	0.15625
10	0.1345	0.140625

23

Using Metals in Building

INTRODUCTION

When ferrous and nonferrous metals are used in the construction of buildings there are many important conditions which must be considered such as strength, fire protection, resistance to corrosion, galvanic action with other metals, staining, and most important, where and why they are used. This unit will cover the uses of iron, steel, stainless steel, aluminum, brass, bronze, copper, and zinc in building construction under the following categories: structural steel, hollow metal, miscellaneous metal, ornamental metal, curtain walls, metal partitions, windows, doors, and flashing.

TECHNICAL INFORMATION

Structural Steel Framing

The structural system of a building may be defined as the method used to support the building both vertically and horizontally. When masonry walls or piers are used for the vertical support, the method is known as wall bearing (see Figure 23-1).

When a building is supported entirely by steel, this is known as

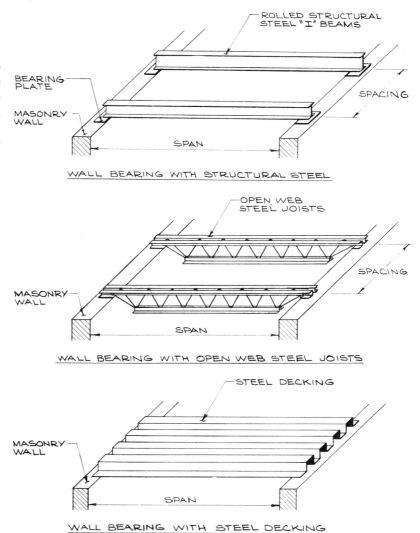

Figure 23-1. Wall Bearing

191

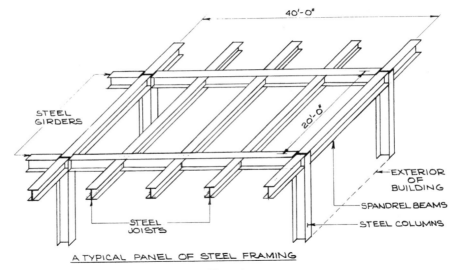

Figure 23-2. *Structural Steel Framing*

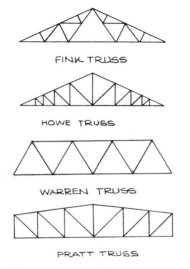

Figure 23-3. *Trusses*

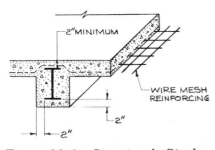

Figure 23-4. *Structural Steel Framing with Reinforced Concrete Slab*

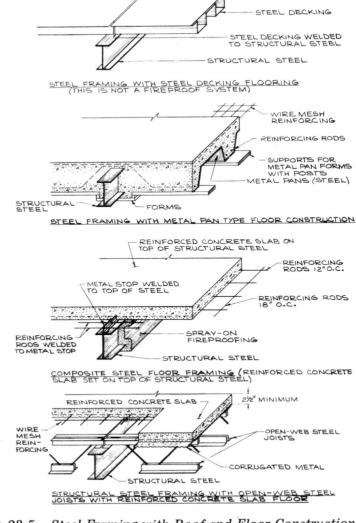

Figure 23-5. *Steel Framing with Roof and Floor Construction*

structural steel framing (see Figure 23-2). When very large areas are to be framed, steel trusses are used either with masonry walls with piers or steel columns (see Figure 23-3).

There are many combinations of steel framing with systems of floor or roof construction — for example, structural steel framing with reinforced concrete slabs, shown in Figure 23-4. Some of the generally used combinations are shown in Figure 23-5.

Hollow Metal

Hollow metal in building construction covers the light-gauge ferrous and nonferrous metals which are bent, extruded, and welded into shapes to form doors, bucks, louvres, partitions, windows, etc. This type of metal work covers such a wide variety of types, shapes, and details that it is impossible to cover them all and therefore only the most typical are illustrated. Interior door bucks and frames are usually made of steel (see Figure 23-6), whereas exterior door bucks and frames are made of steel, stainless steel, aluminum and, in special cases, bronze (see Figures 23-7 and 23-8).

Louvres for interior doors and louvres in exterior walls for heating, ventilation, and air conditioning are generally made of steel or aluminum and, in special conditions, of stainless steel and bronze (see Figure 23-9).

The metals most used in manufacturing windows are steel and aluminum. In special conditions, stainless steel and bronze are used. When using metal windows in building construction it is necessary to consider type, category, surrounding materials, climate, type of glass, orientation, security, insect screens, and most important, type of heating and air conditioning (see Figures 23-10 and 23-11).

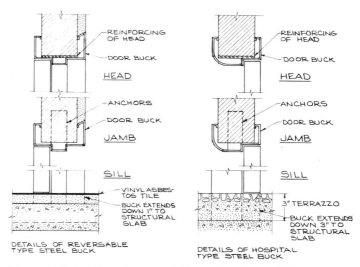

Figure 23-6. *Interior Steel Door Bucks and Frames*

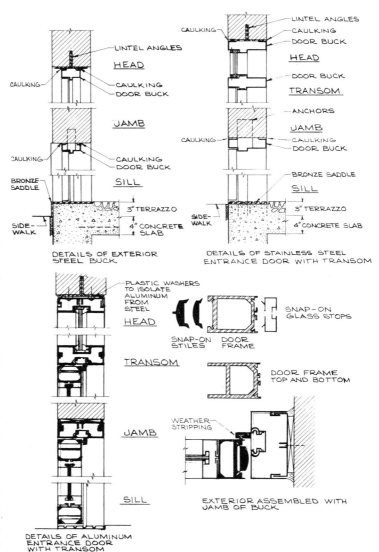

Figure 23-7. *Exterior Steel, Stainless Steel, and Aluminum Door Bucks and Frames*

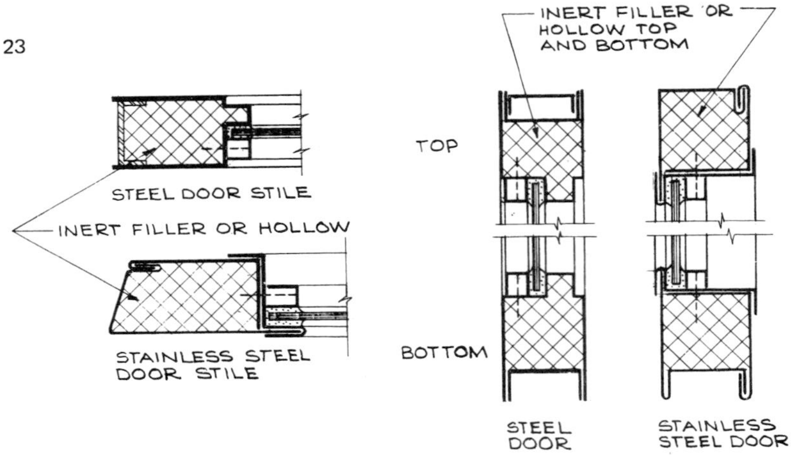

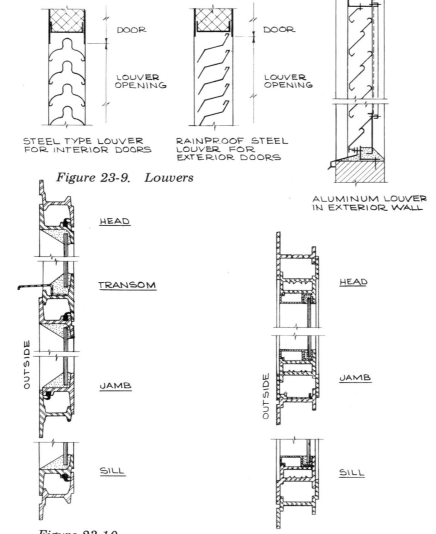

Figure 23-8. *Details of Steel and Stainless Steel Doors*

Figure 23-9. Louvers

Figure 23-10. Aluminum Window Detail

Figure 23-11. Steel Window Detail

Miscellaneous Metal

The ferrous metals — steel, iron and cast iron — are generally used in miscellaneous metal work in building construction and include loose lintels, railings, fences, gratings, ladders, stairs, etc. Some non-ferrous metals are also used in this type of work (see Figures 23-12 through 23-17).

Ornamental Work

Aluminum, bronze, brass, and stainless steel are the metals used for ornamental metal work in building construction, and include grilles, screens, railings, letters, hardware, plaques, etc. (see Figures 23-18 through 23-21).

Curtain Walls

There are four basic types of curtain walls: facing type, prefabricated facing type, grid type, and structural type.

Facing. The facing type consists of a protective finish surface applied to a masonry wall. This surface can be steel with a baked enamel finish

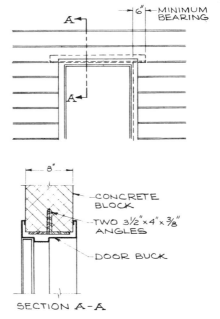

Figure 23-12. Detail of Loose Lintel

(porcelain enamel metal), aluminum, or stainless steel (see Figures 23-22 and 23-23).

Prefabricated. The prefabricated panels consist of an exterior face and a finished interior face with insulation placed between. This type of panel may be as thin as 1-1/2 in. and have insulating characteristics better than a wall consisting of 4-in. brick, 6-in. concrete block, and 5/8 to 3/4-in. plaster

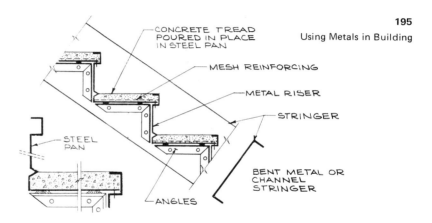

Figure 23-13. Typical Steel Pan Stairs

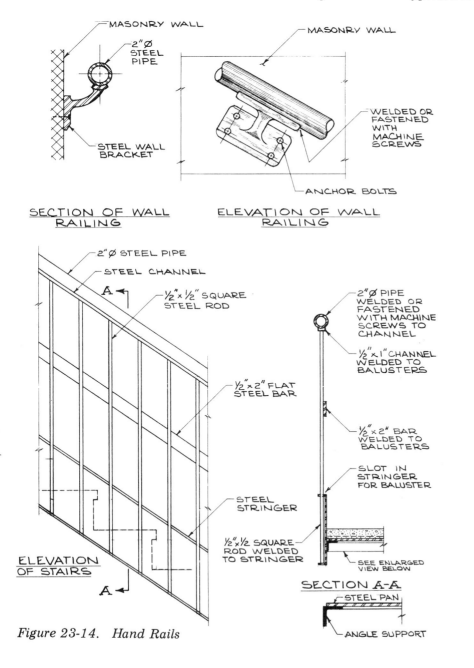

SECTION OF WALL RAILING

ELEVATION OF WALL RAILING

Figure 23-14. Hand Rails

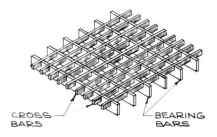

Figure 23-15. Gratings

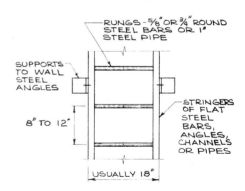

Figure 23-16. Typical Ladder

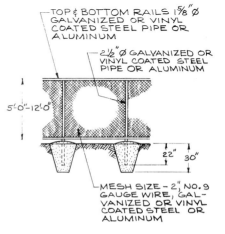

Figure 23-17. Typical Chain Link Fence

finish on the interior (see Figures 23-24 through 23-27).

Grid. The grid system is most often used for stone veneers. (see Figure 17-14).

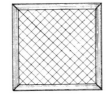
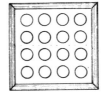

Figure 23-18. Ornamental Grills

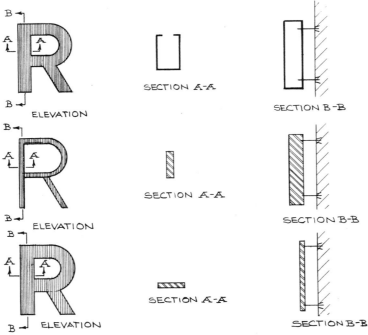

Figure 23-19. Ornamental Screens and Grills

Structural. This type of curtain wall is generally used for buildings of 2 to 4 floors, because the supporting grid becomes too complicated structurally for greater heights (see Figure 23-28).

Advantages of curtain wall type of construction

1. Less thickness of exterior walls;
2. Lightweight, saves steel tonnage of structural steel;
3. Saves time in construction;
4. Because of thinness, valuable rentable floor area is added;
5. When foam insulation is used, heating and air conditioning losses are minimized.

Flashing

Flashing may be defined as both the method and the material used to weatherproof and waterproof the joints wherever different parts of a building or different materials are brought together. Thus, flashing stops the penetrations of water into the interior of the building.

There are two methods of installing flashing: exposed flashing, and concealed flashing. For exposed flashing, the nonferrous metals —namely, aluminum, zinc, and

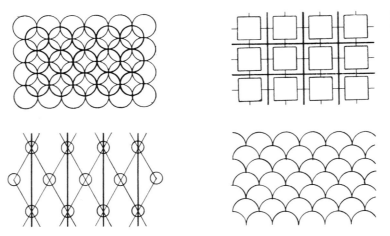

Figure 23-20. Letters

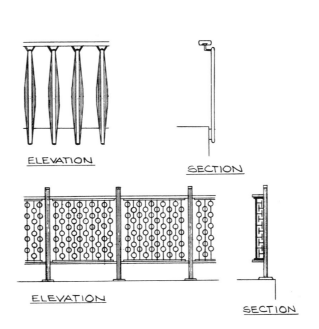

Figure 23-21. Ornamental Railings

lead — are generally used, as is the ferrous metal, stainless steel.

For concealed flashing, not only are the nonferrous metals — aluminum, copper, zinc, and lead — and the ferrous metal — stainless steel — used also coated fabrics, plastics, and coated paper pulp products (see Figure 23-29).

When pipes, ducts, sky domes, railing supports, chimneys, or any other penetration of the roofing material occur, flashing is required (see Figures 23-30 through 23-34).

The edge of roof overhangs require special flashing details in order to terminate the roofing and protect the edge of the overhang (see Figure 23-35).

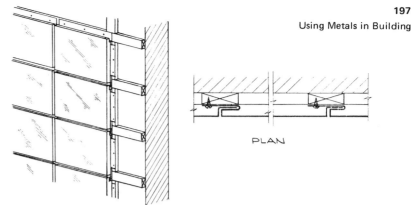

Figure 23-22. Porcelain Enamel Metal Facing Panels

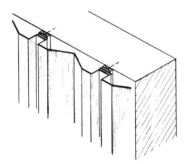

Figure 23-23. Aluminum or Stainless Steel Facing Panels

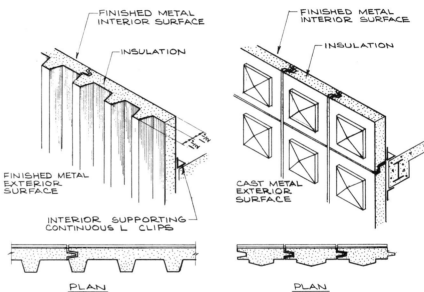

Figure 23-24. Prefabricated Panels

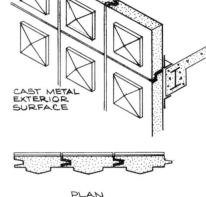

Figure 23-25. Prefabricated Panels (cont.)

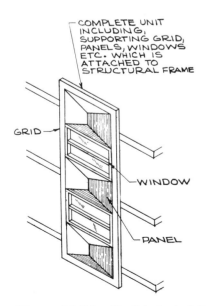

Figure 23-26. Prefabricated Panel

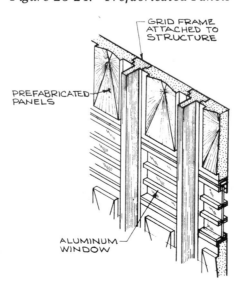

Figure 23-27. Grid System with Panels

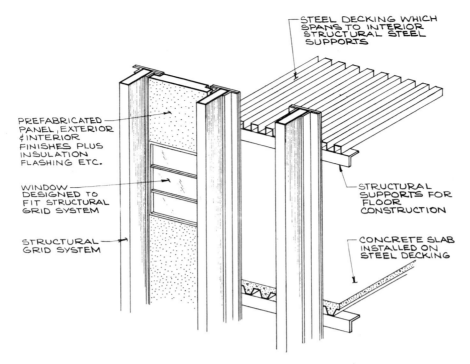

Figure 23-28.   Structural Curtain Wall

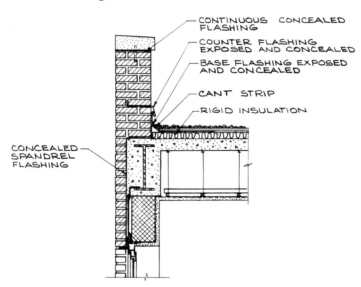

Figure 23-29.   Typical Section Showing Both Exposed and Concealed Flashing

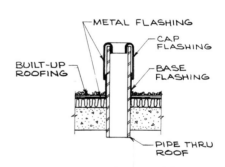

Figure 23-30.   Detail of Pipe through Roof

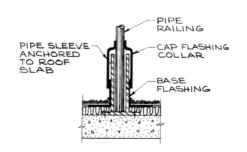

Figure 23-31.   Detail of Pipe Rail Supports

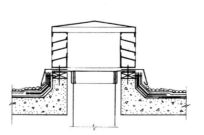

Figure 23-32.   Duct through Roof

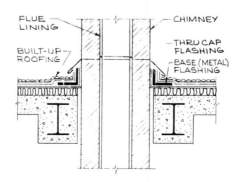

Figure 23.33.   Detail of Chimney

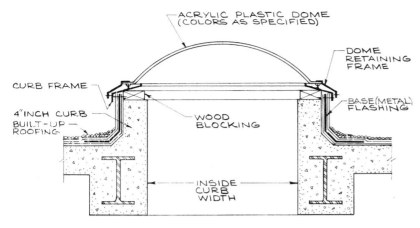

Figure 23-34. Detail of Skydome

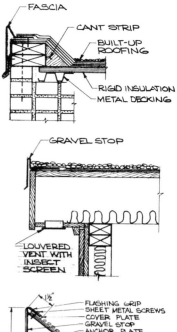

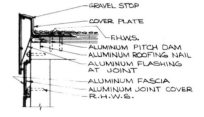

Figure 23-35. Aluminum Gravel Stops and Fasciae

REVIEW EXAMINATION

1. Name four categories of uses of metals in building construction.

2. A wall bearing structural system of a building has what kind of vertical supporting walls?

3. Structural steel framing uses what for all supporting members?

4. Name three (3) combinations of steel framing.

5. Hollow metal work consists of bent, extruded, and welded metals of what sorts?

6. Miscellaneous metal work includes what four things?

7. Aluminum or stainless steel letters used in building construction are included in what category of uses of metals?

8. Name three (3) types of curtain walls.

9. Three advantages of curtain wall type of construction are what?

10. What are two (2) major types of flashing?

11. What three metals are generally used for exposed flashing?

12. Which ferrous metal is used for flashing?

13. What are the metals generally used for windows?

ASSIGNMENT

1. Make a sketch showing both concealed and exposed flashing.

2. In fifty words or less, describe the important conditions that must be considered in using metals in building construction.

SUPPLEMENTARY INFORMATION

Finish hardware. A minor use of metals in building construction is for finish hardware, which constitutes one of the important problems to be solved in constructing buildings. Finish hardware requires schedules, keying and master keying, correct indication of the swing of doors, and types of finish on hardware. Finish hardware includes hinges (known as butts), locks, push plates, door pulls, kick plates, closers, door stops, etc.

24

Glass

INTRODUCTION

This unit deals with the various types of glass in sheet form including *transparent* glass such as window and heavy sheet, plate, insulating, and tempered; and *translucent* glass such as patterned, corrugated, and glass block. Other types of glass materials are also discussed.

TECHNICAL INFORMATION

Originally glass was used for tableware and art objects and in small-sized sheets for windows. Today it has been developed not only into a very wide range of transparent, translucent, and opaque types but also into a wide variety of materials varying from glass fibers to aggregates for concrete and asphalt. Most transparent types of glass transmit 85 to 95% of visible light and are impervious to ultra-violet light.

Definition

Glass is a ceramic material made from inorganic materials — sand (silicon dioxide), soda (sodium oxide), and lime (calcium oxide), with small amounts of aluminum, iron, magnesium, and potassium oxides. Glass can be blown, cast, pressed, rolled, extruded, spun, and ground.

Window and Heavy Sheet Glass

The manufacturing process for window and heavy sheet glass causes distortions that run in one direction through the glass. The degree of these distortions controls the grading of the glass. All window and heavy sheet glass is graded into three qualities:

AA — Specially selected quality
A — Select quality
B — General quality

Window glass is further subdivided by thickness: *single strength*, 0.085 to 0.10 in. thick, with a maximum size of 12 sq. ft., and *double strength*, 0.115 to 0.133 in. thick, with a maximum size of 33.33 sq. ft.
There are two other types of window glass: double strength *greenhouse* quality, which has a slight greenish cast and is available up to a maximum size of $1'-8'' \times 1'-8''$, and *picture*, a thin

glass used for framing, made in 3/64 and 5/64-in. thickness and available in AA, A, and B qualities.

Plate Glass

Plate glass generally has both sides ground and polished so that 90% of total visible light and 77.4% of total radiant energy are transmitted. See Table 24-1 for available thicknesses, maximum sizes, and weights.

Plate glass is used in the manufacture of insulating, heat-absorbing and glare-reducing, and tempered glass; it also is used for mirrors and chalkboards.

When installing plate glass it is important to allow clearances between the glass and the frame in which it is installed (see Table 24-2).

Patterned Glass

Patterned glass (see Figure 24-1) is manufactured in a wide variety of patterns, textures, and designs either on one or on both sides; also, it is available in various degrees of transparency (see Table 24-3).

Patterned glass is manufactured in thicknesses of 1/8, 7/32, and 3/8 in.; in widths of 4'- 0", 4'- 6", and 5'- 0"; and lengths of 8'- 4", 11'- 0", 11'- 4", and 12'- 0".

Insulating Glass

This type consists of two or more sheets of plate glass, or double-strength window glass separated by an air space or spaces. The sheets of glass are joined by a metal-to-glass seal or by the glass itself (see Figure 24-2). It is made in exact sizes and cannot be cut or changed once it is manufactured. When installed, insulating glass should never be in direct contact with the frame and no part of the glass should be covered with paint.

Heat transmission coefficients. Insulating glass has a heat transmission coefficient (U-factor) ranging from 0.47 to 0.67, whereas an ordinary piece of window glass has a U-factor of 1.12. By multiplying the U-factor by the area of the glass by the inside-outside design temperature difference, the heat loss through the glass is found in BTUs per hour.

Example. An outside design temperature of 0° F. and an inside

Table 24-1. Types of Plate Glass

Types of Plate Glass	Thickness	Maximum Sizes	Weight
Silvering, mirror glazing	1/8" to 5/16"	10'-0" × 20'-0"	1.65 to 3.29 lbs. per sq. ft.
Commercial quality	5/16" to 1-1/4"	10'-0" × 25'-0"	4.06 to 16.45 lbs. per sq. ft.

Table 24-2. Minimum Clearances for Installing Plate Glass

Type of Plate Glass	Thickness	Around Edges	Between Glass and Back of Rabbet	Depth of Rabbet
Plate	1/8"	1/8"	1/16"	3/8"
	1/4"	1/4"	1/8"	5/8"
Heavy plate	5/16" and over	1/4"	1/8"	5/8"

FLUTED   RIBBED   HAMMERED

Figure 24-1. Some Types of Patterned Glass

Table 24-3. Methods of Manufacturing Patterned Glass

Finish	Method	How Applied
Fire	Rolling	Both sides
Frosted	Etched with acid	Both sides
Sand-blasted	Compressed air	One side
Textured	Special type roller	One side

design temperature of 70° (winter range). A 100° F. outside temperature and an inside temperature of 70° equals a design temperature of 30° (common range).

A BTU (British Thermal Unit) is an amount of heat that can raise the temperature of 1 lb. of water 1° F. at sea level. Therefore a sheet of ordinary window glass of 12 sq. ft. (assumed temp. diff. 70°) has a heat loss of

$$12 \times 1.12 \times 70° = 940.8 \text{ BTU/hr.}$$

Similarly, a sheet of insulating glass of 12 sq. ft. has a heat loss of

$$12 \times 0.47 \times 70° = 394.8 \text{ BTU/hr.}$$

Heat loss through insulating glass is approximately 42% that of heat transmission through a piece of ordinary window glass of the same size. Insulating glass is available in various thicknesses and air-space widths between the glass sheets. Their relative heat transmission coefficients are given in Table 24-4.

Tempered Glass

This type of glass is manufactured from plate glass and some types of patterned glass. The glass is heated to just below melting point and then the outside surfaces are immediately cooled by jets of air. The result is that the exterior surface is in compression and the interior is in tension and these stresses are in balance.

Tempered glass is three to five times stronger to impact forces than plate glass of the same thickness; therefore it is used in areas where impact and shock are present, such as in doors, sidelights to doors, windows at grade, store fronts, etc. Tempered glass cannot be cut, drilled, or sand-blasted after it is manufactured (see Table 24-5).

When tempered glass is used as a door, all hardware, holes and sandblasting become an integral part of the door in the process of manufacture (see Figure 24-3).

These types of glass absorb high percentages of the total radiant

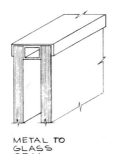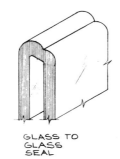

METAL TO GLASS SEAL

GLASS TO GLASS SEAL

Figure 24-2. Glass Seals

Table 24-4. Heat Transmission Coefficient (U-factor) for Various Types of Insulating Glass

Insulating Glass	Thickness of Glass	Air Space			U-factor
Metal seal	1/8"	1/4"			0.64
and one			1/2"		0.57
air space	1/4"			1/4"	0.62
				1/2"	0.52
Metal seal and two air spaces	1/4"		1/4"		0.47
Glass seal and one air space	1/8"		5/16"		0.67

Table 24-5. Types of Tempered Glass

Type of Glass	Thickness	Maximum Sizes	Weight
Plate	1/4" to 1 1/4"	6'–5" × 9'–2"	3.29 to 16.45 lbs. per sq. ft.
Patterned	7/32" to 3/8"	6'–0" × 12'–0"	2.75 to 5.00 lbs. per sq. ft.

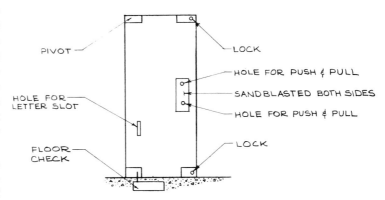

Figure 24-3. Tempered Glass Door Showing Hardware Holes and Sandblasting

heat of the sun and therefore the edges of the glass that are covered by the frame should be of a minimum depth. The reason is that the main area of the glass heats up quickly while the edges of the glass remain cold, and cracking will occur if the frame covers too much of the glass.

Table 24-6. Solar Radiation Transmission

Type of Glass	% of Ultraviolet	% of Daylight	% of Solar Radiation
1/4" plate	67.8	89.1	79.9
1/4" heat-absorbing	44.9	74.7	46.3
1/4" glare-reducing	39.0	40.0	45.0

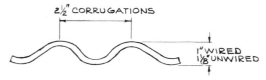

Figure 24-4. Corrugated Glass

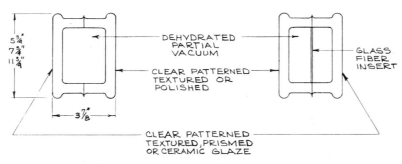

Figure 24-5. Glass Block

Table 24-7. U-Factor and Light Transmission

Type of Glass	U-factor	Light Transmission
Window glass	1.12	85% to 95%
Glass block, hollow	0.44 to 0.60	0% to 85%

The glare-reducing and heat-absorbing qualities are developed by the chemical composition of the glass. These types of glass are made in plate, heavy sheet, patterned, and tempered; in the plate-type federal specifications require that not more than 50% of total solar radiation be transmitted (see Table 24-6).

Wire Glass

Hexagonal or diamond-shaped wire mesh is embedded in the middle of plate, patterned, or corrugated glass in the manufacturing process. Wire glass is 1/4 in. or thicker, with the wire mesh not larger than 7/8 in., and the gauge of the wire not less than No. 24.

Wire glass is used (1) where fire resistance is required (size limited to 5 sq. ft. with a maximum length of 4'-0" or a maximum width of 4'-0"), and (2) in areas where impact and abuse are present and where flying broken glass would be dangerous, such as skylights, overhead lighting, etc.

Corrugated Glass

This is 3/8-in. thick rolled glass with a pattern on both sides, translucent, wired or unwired, with a standard length of 10'-0" and a maximum length of 12'-0" (see Figure 24-4).

Unwired corrugated glass is generally used for decorative partitions, both exterior and interior, where light and privacy are desired.

Wired corrugated glass is used for roofs, skylights, partitions, etc., where breakage must be avoided.

Glass Block

Glass blocks are manufactured in two types, solid and hollow. Hollow glass blocks have a dehydrated, partial-vacuum air space (see Figure 24-5) which serves an insulating function. The difference in the heat transmission coefficient (U-factor) and in light transmission between window glass and glass block is shown in Table 24-7. Solid glass blocks generally are used for decorative panels and in sidewalks for cellar or basement lights.

Glass block should never be used for load-bearing. When installed in a frame, the maximum area is 144 sq. ft. with a maximum height of 20'-0" or a maximum length of 25'-0". When installed without a frame, the maximum area is 100 sq. ft. with a maximum height of 10'-0" or a maximum length of 10'-0".

The minimum clearances for hollow glass blocks in an opening without a frame are: head — 1/2 in., jamb — 3/8 in., and sill — 1/4 in. The minimum clearance in an opening with a frame are: head, jamb, and sill — 1/4 in.

Other Types of Glass Materials

There are numerous other types of glass material with architectural uses and characteristics (see Table 24-8).

Table 24-8. Glass Materials Used in Architecture

Material	Characteristics	Architectural Use
Cellular glass	Black, solid with tiny trapped air bubbles, with a U-factor of 0.13 to 0.09	Roof and wall insulation; industrial-type insulation
Chalkboards	Tempered plate glass with a colored vitreous glaze on one side	Chalkboards in schools, colleges, and universities
Glass fibers	5 to 16 microns thick	Insulating materials; reinforcing for plastics, papers, and textiles
Granular glass	Broken glass or waste glass, heated and made into aggregates	Aggregates for concrete
Heat-strengthened glass	Black to white in color and with ceramic glazes	Spandrels and curtain wall construction
Mirrors	Plate, window, and picture glass with one surface treated with a reflecting, thin coat of metal	Mirrors and one-way mirrors
Powdered glass	Finely ground glass power	Filler for resilient flooring, paint, asphalt roofing and siding
Structural glass	Black, white, and various colors	Exterior and interior facing for buildings; toilet partitions
Laminated glass	Sheets of window or plate glass with clear plastic placed between	Bulletproof glass; areas where severe shock may be encountered

REVIEW EXAMINATION

1. What is glass made from?

2. What are the three qualities of window and heavy sheet glass?

3. Is plate glass ground and polished on one side or both sides?

4. Name three methods by which patterned glass is made.

5. How much stronger is tempered glass than plate glass?

6. How is insulating glass sealed?

7. A piece of insulating glass has approximately what percent of the heat loss per hour that window glass of the same size has.

8. Glare-reducing and heat-absorbing glass absorb a high percentage of what?

9. In wire glass, the mesh size should not be larger than what and the gauge of the wire should not be less than what?

10. Does corrugated glass have a pattern on one side or both sides?

11. Where is solid glass block used?

12. When glass block is installed in a frame, what should its maximum area be?

ASSIGNMENT

1. Make an elevation scale drawing of a tempered glass door showing all the hardware, holes, and sand blasting which become an integral part of manufacturing the door.

2. In fifty words or less explain how to obtain the BTU/hr. heat loss of a sheet of insulating glass, 24 sq. ft. with a U-factor of 0.67 with an outside temperature of 0° F. and an interior temperature of 70° F.

Table 24-9. Maximum Range of Glass Sizes in Sq. Ft. for a Given Wind Velocity

Wind Velocity	Window Glass Thickness		Plate Glass Thickness			
	0.085"	0.133"	1.8"	1/4"	1/2"	1 1/4"
30 mph	35 sq. ft.	64.5 sq. ft.	72 sq. ft.	288 sq. ft.		
40 mph	17.5 sq. ft.	32.25 sq. ft.	36 sq. ft.	144 sq. ft.		
65 mph	8.7 sq. ft.	16.1 sq. ft.	18 sq. ft.	72 sq. ft.	244 sq. ft.	
120 mph	2.5 sq. ft.	4.6 sq. ft.	5 sq. ft.	20 sq. ft.	82 sq. ft.	81 sq. ft.

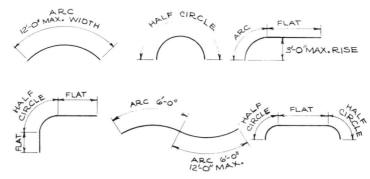

Figure 24-6. Bent Glass

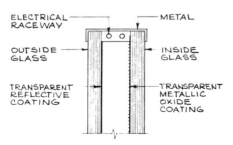

Figure 24-7. Insulating Glass

Table 24-10. Materials Used for Coloring Glass

Material	Color
Copper	Blue-green, ruby
Cadmium	Ruby, orange, yellow
Iron	Yellow-green, blue-green, amber
Gold	Ruby
Manganese	Pink
Nickel	Brown, purple
Cobalt	Blue

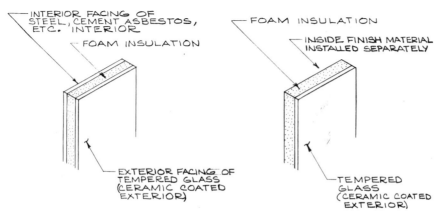

Figure 24-8. Prefabricated Glass Curtain Wall Panels

Limiting Factors

Effect of wind velocity. The size of glass used in glazing windows, doors, and store fronts is controlled by the velocity of the wind (see Table 24-9).

Bending. Glass may be bent within definite limitations, but the maximum rise is 3'-0" and maximum width is 12'-0" (see Figure 24-6).

Colored Glass

Glass is manufactured in a variety of colors (see Table 24-10).

Insulating Glass as Heating Unit

Insulating glass (see Figure 24-7) is now manufactured using the inner sheet of glass as an electrical heating element. The outside sheet of glass is transparent with a reflective coating, but the inside sheet has a transparent electrical-resistant metallic coating on the air-space side, with electrical connections. The metallic oxide coating heats the glass and thus eliminates condensation and cold drafts, and there is no BTU loss through the window.

Glass Used on Curtain Wall Panels

Tempered glass with insulation or with insulation and an interior facing are prefabricated for use in curtain wall panel construction (see Figure 24-8).

Reflective Glass

This type of glass consists of a flat glass (clear or tinted) coated on the outside with a transparent metallic oxide coating. This provides the glass with a permanent, durable, light- and heat-reflective exterior surface that has mirrorlike qualities. The outdoor reflectance is 35% and visible transmittance varies from 18% to 39%.

1. What is the cut stone or dimension stone most likely to be used for stair treads and risers?

2. The mortars used for setting stone, pointing and grouting are made with what kind of portland cement.

3. In most veneer stonework the back-up material should be what? Why must the space behind the veneer be open?

4. A stone measuring 6 sq. ft. and 7/8 in. thick requires how many anchors when used as a veneer?

5. What are the two materials used in setting floor and wall tile?

6. When installing large floor areas with tile, what must be installed to avoid expansion and contraction of the structural slab that may cause cracks in the setting bed and tile joints?

7. Wood from the evergreens, or coniferous cone-bearing trees, belong to what family of trees?

8. What is all lumber called when first cut from trees?

9. Lumber may be seasoned by what two drying methods?

10. The American Lumber Standards Association specify that lumber be classified according to what principle uses?

11. What is the actual size of a 2 × 4-in. piece of lumber?

12. Yard lumber is used for what building purposes?

13. What two main divisions are there in yard lumber on the basis of quality?

14. What is D & M lumber?

15. What causes decay in wood?

16. What is an oil-type preservative for wood?

17. What is plywood made of softwood used for?

18. What is plywood made of hardwood used for?

19. What are the two main types of plywood that are manufactured?

20. What is the common and basic size of the plywood panel?

21. When plywood is used for exterior siding, what is the minimum thickness required?

22. What is the acceptable thickness of plywood used as exterior wall sheathing when the studs are 16″ OC?

23. For what purpose is particle or chip board used with high-pressure plastic laminate veneers and hardwood veneers?

24. In what thicknesses is laminated fiber board decking available?

25. What is the thickness of the smaller pieces of lumber that are laminated together to form a larger and stronger member?

26. Solid timbers are usually cheaper than laminated sections, except where extremely heavy loads are to be supported, for spans up to what dimension?

27. Laminated timbers are generally manufactured to what maximum and in what increments?

28. What are the three specific types of three-hinged arches?

29. What is another name for the conventional Wood Frame Western construction?

30. The studs of what two kinds of framing rest on the sill?

31. What advantage does post and beam construction provide in interior room planning?

32. What is the diameter of an 8d and 9d nail?

33. What kind of action begins when different metals or alloys are joined together with the presence of moisture?

34. What is the specific gravity of aluminum?

35. What is the specific gravity of copper?

36. Metals are either ferrous or nonferrous — which is zinc?

37. In the field of construction, dampers, manhole covers, gratings, pipe, plumbing fixtures are made of what material?

38. In building construction, what is the framing that consists of various rolled sections?

39. What is copper combined with to make bronze?

40. What bronze-colored metal is now known as architectural bronze?

41. How are ferrous and non-ferrous wires designated?

42. What is the type of construction called when the entire method of supporting a building is in steel?

43. What are the two types of flashing?

44. What percentages of visible light are transmitted by most types of transparent glass?

45. What are the basic materials from which glass is made?

46. What is the glass consisting of two sheets of glass separated by an air-space?

47. What is the heat transmission coefficient (U-factor) of insulating glass consisting of three sheets of glass with two air-spaces?

48. A BTU is an amount of heat that can raise the temperature of how much water 1° Fahrenheit at sea level.

49. What type of glass is used where fire resistance is required?

50. What is the U-factor of window glass?

25

Installation of Glass
and Glass Materials

INTRODUCTION

This unit deals with the installation of the various types of glass including transparent, translucent and opaque; with the preparations necessary to the materials and openings that will receive the glass; and with the various glazing materials.

TECHNICAL INFORMATION

Installation of Glass

The installation of transparent and translucent glass into a frame is called "glazing," whereas other types of glass materials require their own special methods of installation. The thickness and type of transparent or translucent glass for any type of installation should be checked for size of opening and wind loads.

Glazing. Glazing includes preparing the frame to receive the glass, allowing clearances between the glass and the frame (glass must never be in tension), making a watertight seal between the glass and the frame with glazing materials, and installing setting blocks, spacers, and glazing points to keep glass completely separated from frame.

Clearance between glass and frame. The various types of glass require different minimum clearances between glass and frame: around edges of glass, faces of glass to back and front of frame, and depth of the slot (rabbet) to hold the glass (see Table 25-1 and Figure 25-1).

Table 25-1. Clearances between Glass and Frame (in inches)

Type of Glass	Thickness	Minimum Clearance: Edges of Glass to Frame	Back and Front Faces of Glass to Frame	Depth of Slot (rabbet)
Window; single- and double-strength	0.087 0.113 1/8	1/8	1/16	3/8
Heavy sheet	3/16 7/32	1/8 1/4	1/8	1/2 3/4
Plate	1/8 1/4 and	1/8 1/4	1/16 1/8	3/8 5/8
Patterned	1/8 3/16 7/32	1/8 1/4	1/8	3/8 1/2 5/8
Insulating	1/2 3/4 1	1/8 1/4	1/8	1/2 3/4
Heat-absorbing and glare-reducing	0.125 1/8 3/16 7/32 1/4 3/8	1/4 min.	1/8	3/4

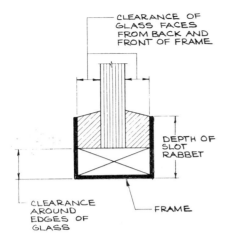

Figure 25-1. Clearance between Glass and Frame

When glass is installed in frames, setting blocks and spacers are used to keep the clearances required of glass to frame (see Figures 25-2 and 25-3).

Preparation for glazing. Preparation of the materials into which the various types of glass are to be installed include the following steps:

1. All surfaces should be thoroughly dry and rough projections removed; installation should not be done when the temperatures are below 40° F.

2. Wood should always be painted with a prime coat of paint.

3. Steel should be treated with a rust-inhibiting type of prime paint.

4. Aluminum should have any shop protective coating removed. Steel frames do not have to be prime painted if they have received a shop-applied bonderized paint or a zinc coating. Aluminum should be prime painted if an oil-based glazing compound is used.

Hardwoods (absorbent) may be prime painted with varnish when an elastic glazing compound is used; hardwoods (nonabsorbent) do not have to be prime painted if a metal sash putty is used.

Glazing materials. There are many types of glazing materials available for making the joint between the glass and frame watertight:

1. *Putty* — white paint solid, linseed oil and with or without drying oils.

2. *Metal sash putty* — made of materials that will adhere to nonporous materials such as metals and nonabsorbent hardwoods.

3. *Elastic glazing compound* — made from processed oils and white paint solids that will remain plastic and resilient over a long period of time.

4. *Compression-type* — extruded or molded shapes of vinyl, neoprene, or other synthetics. In order to achieve a watertight joint the material must be compressed not less than 15%.

5. *Zipper types* — extruded or molded shapes of vinyl, neoprene, or other synthetics which can be compressed or squashed in order to install the glass and then made into a watertight joint by being stabilized by the insertion of a zipper strip (see Figure 25-3).

Corrugated Glass

In installing corrugated glass for roofs and skylights, special care should be taken in lapping, attaching, insuring watertightness, and answering the problems of condensation (see Figure 25-4).

Glass Block

Glass block walls or partitions cannot be used for load-bearing, therefore the frame or opening must fit the dimension of the glass block and the top (head) of the frame or opening must take the load-bearing from the glass block. Glass block installed within an opening of masonry or steel frame is limited in size to 144 sq. ft., with a maximum height dimension of 20'-0" and a maximum width dimension of 25'-0". Glass block partitions or walls without support are limited in size to 100 sq. ft. with a maximum height of 10'-0" and a maximum length of 10'-0". In order to allow for clearances, expansion, and possible deflection, it is necessary to use expansion strips, nonhardening caulking compound, oakum, and asphalt strips to break bond as shown in Figure 25-5.

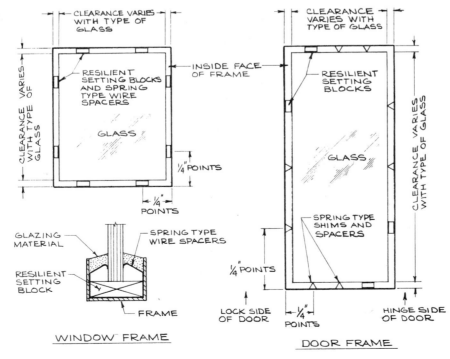

Figure 25-2. Window and Door Frames

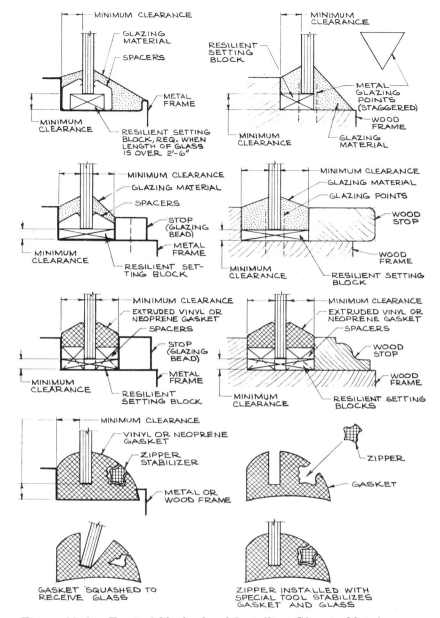

Figure 25-3. *Typical Methods of Installing Glass in Metal or Wood Frames*

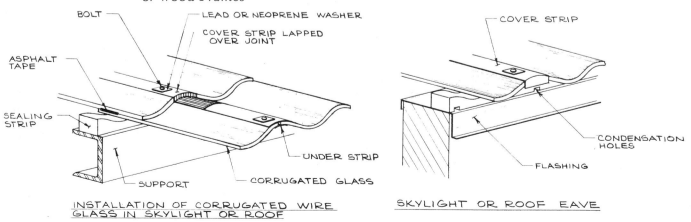

INSTALLATION OF CORRUGATED WIRE GLASS IN SKYLIGHT OR ROOF

SKYLIGHT OR ROOF EAVE

Figure 25-4. *Corrugated Glass Skylights and Roofs*

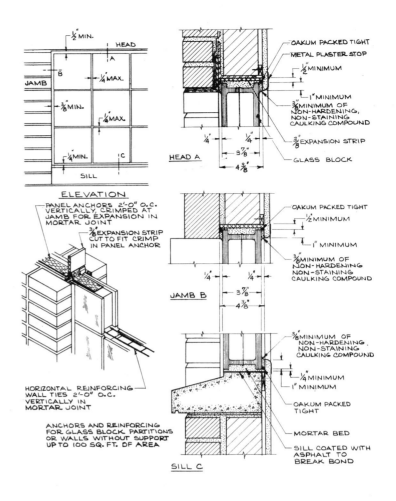

Figure 25-5. Details Showing Installation of Glass Blocks

REVIEW EXAMINATION

1. The installation of glass is called what?

2. What is used to keep the glass face free from the frame?

3. To make a watertight seal between glass and frame what material is used?

4. In installing glass, where are the setting blocks located?

5. How should the frame be prepared for glazing?

6. Glass should be installed above what temperature?

7. When an oil-based glazing compound is used, what should be done to the aluminum frame?

8. Can glass block be used for load-bearing?

9. What is used when glass block is installed to allow for clearance, expansion, and deflection?

10. When installing corrugated glass for roofs and skylights, name three areas where special care should be taken.

ASSIGNMENT

1. Sketch at least four details showing the installation of glass in wood and metal frames.

Other Types of Glazing Materials

For special type of glass materials there are available two other types of glazing materials:

1. *Polybutene tape*, a nondrying mastic made in extruded ribbon shapes of various widths and thicknesses, which must be applied with pressure. It has the in-built quality to remain plastic over a long period of time.

2. *Polysulfide elastomer*, a two-part synthetic rubber mixed on the job. This material requires the surrounding surfaces to be carefully protected because when it sets it is almost impossible to remove.

Heat-strengthened Glass Installation

Heat-strengthened glass is manufactured from plate or patterned glass by applying a ceramic glaze to one side and heating until it is fused with the glass. It is opaque, similar to tempered glass, and is generally used in curtain wall construction. This glass has specific minimum clearances, as shown in Figure 25-6.

Structural Glass Installation

Structural glass is a ground and polished colored glass, either tempered or heat-strengthened. It cannot be cut or drilled, therefore shop drawings of the exact sizes of the units must be made before it can be manufactured. Generally it is used for interior and exterior wall surfaces, toilet partitions (see Figure 25-7), and counter tops. In installing, the back-up material must be plaster or a masonry material which has been sealed and made waterproof (see Figure 25-8).

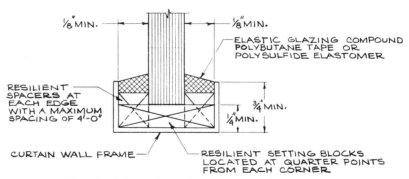

Figure 25-6. *Installing Heat-Strengthened Glass*

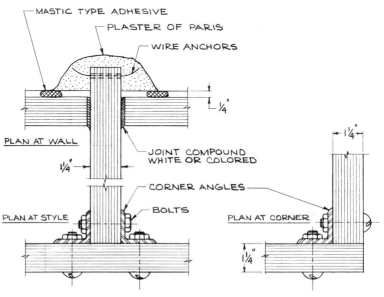

Figure 25-7. *Toilet Partition Details*

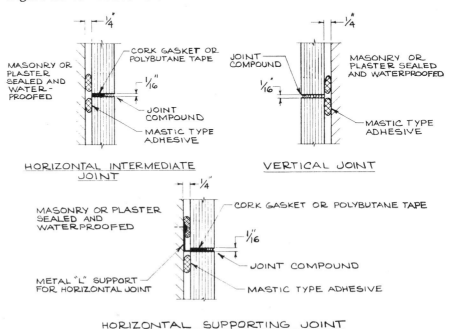

Figure 25-8. *Structural Glass Exterior Walls*

26

Plastics

INTRODUCTION

This unit deals with the synthetic materials used in construction that are grouped under the general term "plastics." They include flooring, sheets (transparent, colored, opaque, flat, corrugated, and deformed), plastic laminates, gaskets, sealants, foam and film, as well as many new products and construction units appearing on the market.

TECHNICAL INFORMATION

The word "plastic" means a synthetic material which is not rubber, glass, wood, natural resin, or metal. The word now covers such a large diversified group of materials that it can only be used in a very broad sense and cannot be used specifically to describe any one particular synthetic material.

Table 26-1 shows the main types and their uses. Plastic products include paints, textiles, insulation, baked enamels, and thousands of small items such as electrical devices, pipe, hardware, bathroom accessories, etc.

Types of Plastics

Synthetic materials can be broadly divided into the following types:

Table 26-1. Types of Plastics

	Plastic	Name*	Major Use
1.	Acrylic	Lucite Plexiglass	Skydomes, glazing, light fixtures
2.	Alkyd	Alkyd	Vehicles in paints, lacquers, and enamels
3.	Allyl	Contact adhesives	Applying plastic laminates and other sheet material to walls and surfaces
4.	Cellulose	Cellophane, Lumerith	Transparent thin sheets, artificial leather
5.	Chloroprene	Neoprene	Gaskets, sealants, and joint fillers
6.	Epoxy	Epoxies	Tough waterproof paints for walls and floors
7.	Ethyl cellulose	Household adhesives	General adhesives
8.	Latex	Latex emulsions	Water-base paints
9.	Melamines	Melamines	Plastic laminates, tableware
10.	Phenol-formaldehyde	Phenolics, Bakelite	Electrical devices, hardware, baked enamels, plastic laminates
11.	Polyamides	Nylon	Rollers, pulleys, fabrics
12.	Polycarbonate	Lexan, Merzon	Glazing
13.	Polyester	Dacron, Mylar	Fabrics; transparent tracing paper, sheets
14.	Polyethylene	Polyethylene	Dampproofing and waterproofing, sheets, pipe
15.	Polysulfide	Thiokol	Sealants, glazing compounds
16.	Polyurethane	Polyurethane, Perlon, Igamid	Insulating material, foam-type insulating material
17.	Styrene	Butadiene	Tapes, glazing compound
18.	Urea-formaldehyde	Ureas	Baked enamels
19.	Vinyl	Vinyl	Resilient flooring, gaskets, joint fillers, paint, lacquers

*Many of these are trade names and are offered as examples. No attempt is made here to be all-inclusive.

217

(1) those derived from cellulose, (2) synthetics resins, (3) those derived from proteins and natural resins, and (4) synthetic rubbers.

Synthetic resins can be divided into two types: *thermosetting*, which are formed (shaped) by heat but cannot be reformed by heat; and *thermoplastic*, which can be softened by heat and upon cooling regain their original properties.

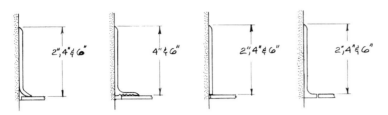

Figure 26-1. Vinyl Bases

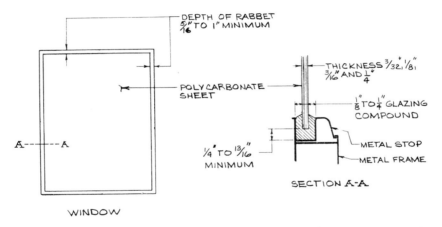

Figure 26-2. Polycarbonate Windows

Table 26-2. *Relation of Sheet Size to Type of Glazing Compound*

Maximum Longest Dimension (in inches)	Type of Glazing Compound
24	Elastic (nonhardening) glazing compound
24 to 96	2-part polysulfide, polybutane, or polyurethane sealant
Greater than 96	Silicone sealants only

Vinyl Flooring

One type of flooring is solid vinyl and is manufactured in rolls and in tile form. Squares are available in 6 × 6, 9 × 9, 12 × 12, and 18 × 18 in.; rectangles in 3 × 9, 3 × 12, and 6 × 12 in.; both in thicknesses of 1/16, 3/32, and 1/8 in. Other sizes and thicknesses are available on special order. Vinyl sheet flooring is manufactured in rolls 6'-0" wide and 50'-0" in length, 0.09 and 0.07 in. thick, and available in a wide variety of colors, textures, and patterns.

A wide variety of vinyl bases are manufactured, as shown in Figure 26-1. They are available in solid vinyl in a limited range of colors.

Vinyl Asbestos Flooring

This type of flooring is manufactured from vinyl, asbestos fibers, and fillers, in squares 9 × 9 and 12 × 12 in., in thicknesses 1/16, 3/32, and 1/8 in. All types of vinyl asbestos flooring are available in a wide variety of colors, textures, and patterns. Feature strips and inserts are also available varying according to manufacturer.

Transparent, Colored, and Opaque Sheets

Polycarbonate sheets are extremely tough, almost unbreakable, and remain stable from 60° to 300° F. Because of these characteristics, polycarbonate sheets have become a substitute for glass in areas where vandalism, thermal shock, and impact are present, and where safety from breakage is required. It also is used in areas where protective mesh, screens, and railings are usually installed. Installation clearance, glazing thickness, and rabbet depth depend on size and thickness of sheet (see Figure 26-2). The type of glazing compound depends on the size of sheet (see Table 26-2). Polycarbonate sheet has a light transmission of 82 to 89%, depending on thickness; it has self-fire-extinguishing characteristics, and it is manufactured in a variety of colors.

Other types of plastic sheet material are manufactured either transparent, colored, or opaque and are used for skydomes and decorative screens and for glazing. They do not have the toughness of polycarbonate sheet but will not break. In general, all these materials are not as hard as glass and their surfaces can be damaged. See Table 26-3 for thickness, dimensions, and glazing information.

Corrugated Plastic Sheets

This material is manufactured with glass fibers used as reinforcing. Translucent and available in limited colors, it finds use in the construction of buildings for porch roofs, shades, screen walls, and clerestory windows. The corrugations and the glass fiber give it the strength to span to a maximum unsupported length of 4'-6" for a 100-lb. load (see Table 26-4).

Plastic Laminates

This type of thin, hard plastic sheet with decorative, colorful designs and patterns and a very durable surface is used in the construction field as finish surface for walls, partitions, doors, countertops, kitchen cabinets, and furniture (see Figure 26-3). It can be applied either in a shop or at the site. It is available in three types: general purpose, cigarette-proof, and special types for forming curves (see Table 26-5).

Gaskets and Sealants

Plastics were adopted very early in the construction field for glazing. With the development of curtain wall type of construction, plastics were used as sealants. Today, we find a tremendous variety of plastic gaskets, tapes, sealants, fillers, etc. Figure 26-4 shows some of these newer methods.

Plastic sealants are available in a form to be applied by a caulking gun as two compounds which are mixed and then applied, and also as

various types of sealing tapes. These plastic sealants all have the follow-

Table 26-3.   Manufactured Sizes

Thickness	1/16", 0.080" and 3/32"
Length	4'-0", 5'-0", 6'-0", 8'-0", 10'-0" and 12'-0"
Width	1'-4", 2'-0", 2'-8" and 4'-0"
Glazing	All standard window glass sizes

Table 26-4.   Data for Corrugated Plastic Sheets

Corrugation Spacing	Thickness	Length	Width	Maximum Span for 100-lb. Load
1/4"	1/16"	8'-0", 10'-0", 12'-0"	2'-2", 3'-0"	2'-8"
2 1/2"	1/16"	4'-0", 5'-0", 6'-0", 8'-0", 10'-0", 12'-0"	2'-2", 2'-10", 3'-4", 4'-2"	4'-0"
2.67"	1/16"	8'-0", 10'-0", 12'-0"	2'-11"	4'-6"
3"	3/32"	8'-0", 10'-0", 12'-0"	2'-6"	4'-6"
4.2"	3/32"	8'-0", 10'-0", 12'-0"	3'-6"	4'-6"

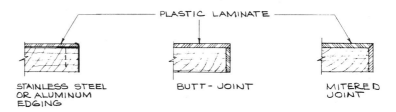

Figure 26-3.   Methods of Treating Plastic Laminates at Edges

Table 26-5.   Types and Sizes of Plastic Laminates

Type	Thickness	Width	Length
General purpose	1/16"	2'-0", 2'-6", 3'-0"	5'-0", 6'-0" 7'-0", 8'-0", 10'-0"
Cigarette-proof			
For framing radii and curves	1/20"	2'-6", 3'-4", 4'-0"	6'-0", 8'-0", 10'-0"
Cigarette-proof			

ing characteristics: they remain adhesive, flexible, and elastic for ten to twenty years; they are used for glazing, caulking, and as fillers for all types of joints between different materials. Plastics have answered the problem of construction joints in masonry and in expansion joints in large buildings (see Figure 26-5).

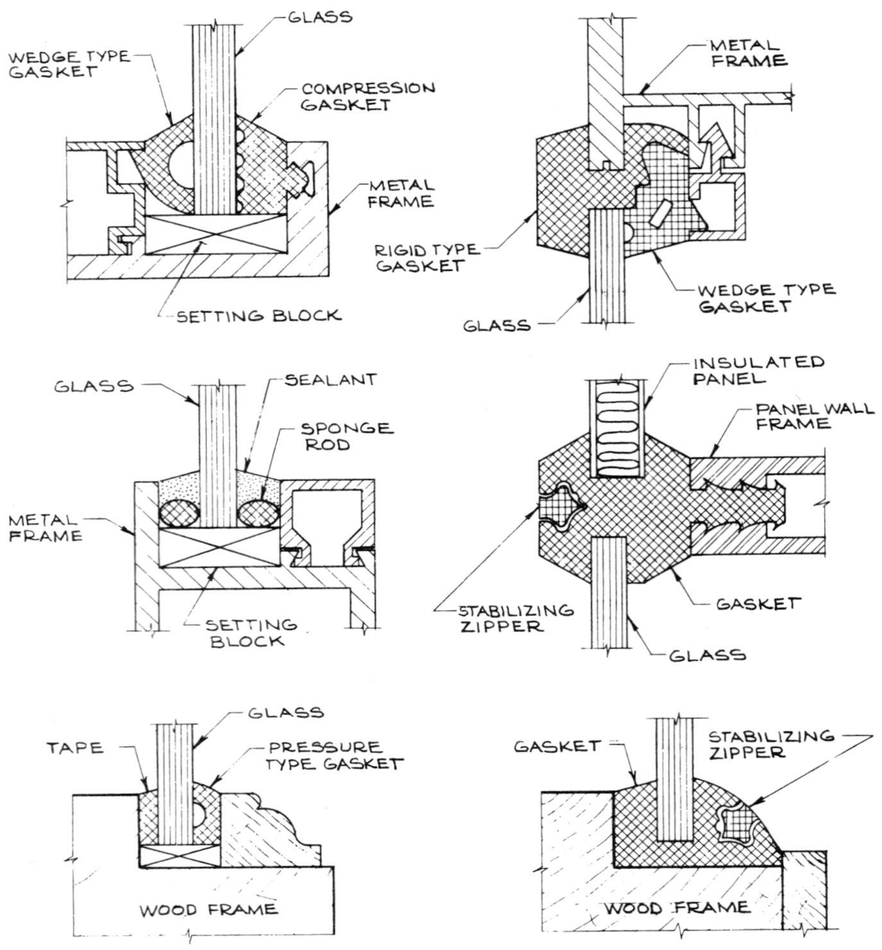

Foam and Film Plastics

Polyurethane and polystyrene plastics produce two types of insulation, *pre-formed* and *foamed*. A similar type of pre-formed plastic insulation is styrofoam.

Pre-formed plastic insulation is manufactured in sheets and in block form. Because it consists of thousands of sealed air bubbles, it has exceptional U-factors (1 in. has a U-factor of 0.12). It is used for all types of insulation and has the added advantage of serving as a moisture and vapor barrier.

The *foam type* answers the problem of insulation of areas that in the past could never be adequately insulated with other types of insulation. When the ingredients are mixed and injected into these areas, their foaming action completely fills the voids.

Another by-product of this foam action is that with heat, certain of the polyurethane plastics become adhesive. Through this method, sandwich panels can be manufactured (see Figure 26-6). With other types of foam plastic the finish surfaces can be bonded to the foam with special types of adhesives. Any combination of materials can thus be used for the outside and inside surfaces — metal, plywood, wood, cement, asbestos, etc. — to manufacture a sandwich panel. These panels of metal and 1-1/2 in. of foam insulation have better insulation value (U-factor 0.102) than 8 in. of brick with 1-1/2-in. metal furring and lath and plaster (U-factor 0.32) or the normal 4-in. stud partition with sheathing, siding, 2 in. of insulation and plaster board (U-factor 0.11).

Plastic film finds many diversified uses — for example, as tracing paper (mylar), as underlayment beneath concrete slabs on earth for moistureproofing, as waterproof membranes for bathrooms, and as flashing and roofing material.

Figure 26-4. Installation of Glass Using Plastics

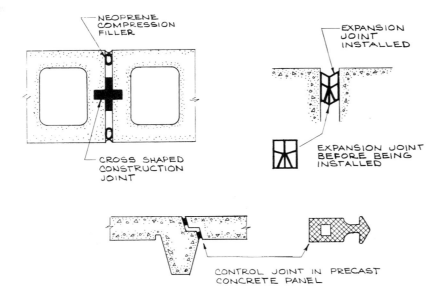

Figure 26-5. Construction Joints in Masonry

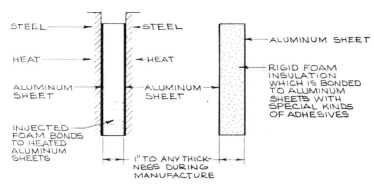

STEEL — STEEL

HEAT — HEAT

ALUMINUM SHEET — ALUMINUM SHEET

INJECTED FOAM BONDS TO HEATED ALUMINUM SHEETS

1" TO ANY THICK-NESS DURING MANUFACTURE

ALUMINUM SHEET

RIGID FOAM INSULATION WHICH IS BONDED TO ALUMINUM SHEETS WITH SPECIAL KINDS OF ADHESIVES

Figure 26-6. Foam Insulating Panels

REVIEW EXAMINATION

1. Can the word "plastic" be used to describe plastic laminate?

2. Can a thermosetting plastic be re-formed by heat?

3. Vinyl tile flooring is manufactured in what thicknesses?

4. Vinyl asbestos tile is available in what sizes?

5. What are the advantages of using plastic transparent sheet in place of glass?

6. Are plastic sheet materials as hard as glass?

7. Name three uses of corrugated plastic sheet material.

8. What are the two types of plastic laminates?

9. Where is a plastic tape used?

10. Are plastic materials used for masonry construction joints?

11. With heat, do foam plastics have adhesive qualities?

12. Name three uses of plastic film.

ASSIGNMENT

1. Make a scaled drawing showing three methods of edge treatment of plastic laminates.

2. Make a scaled drawing showing two methods of installing glass with plastic materials.

SUPPLEMENTARY INFORMATION

Plastic Laminates

These are best known under trade names such as Formica, etc. They are manufactured of several layers of kraft paper impregnated with phenolics, a patterned or colored sheet saturated with melamine and a hard, transparent finish surface of melamine. The layers are then exposed to intense heat and pressure to form the finished plastic laminate sheet. In order to make them cigarette-proof, a layer of aluminum foil is added to dissipate the heat. All plastic laminates are resistant to heat up to 275° F.

Thermosetting (thermocuring) plastics have a molecular structure in which the molecules are linked in a three-dimensional arrangement.

Thermoplastic plastics have a molecular structure in which the molecules are linked in a linear arrangement.

Fashion Trends in Flooring

The manufacturers of vinyl and vinyl asbestos tiles and sheet vinyl are constantly creating new colors, textures, and patterns to answer consumer demands. Care should be taken that the latest and current colors, textures, and patterns chosen, are *not* a discontinued product.

27

Paints and their Application

INTRODUCTION

Originally, paint consisted of a natural oil (linseed oil) as a vehicle to which finely ground white solid particles of white lead (basic lead carbonate), known as pigment, were added. When properly mixed, the white lead solids were held in suspension in the vehicle.

Painting consisted of applying the paint by brush to a surface. As the applied paint dried, a solid protective film developed. To obtain a good paint surface, it was necessary to apply from three coats (prime coat, undercoat, and finished coat) to six coats (prime coat, second prime coat, undercoat, second undercoat, under-finish coat, and finish coat). The white paint was colored by adding color pigments; it could be thinned by adding a solvent or thinner; and it could be removed by applying a solvent such as turpentine.

The development of synthetics (plastics) and a variety of white paint solids has brought about a complete change in paints and painting procedures. With new research and development, this change in paints and techniques is still continuing. This unit will explore paints, resins, driers, solvents, extenders, and other types of paint and painting.

TECHNICAL INFORMATION

Paint

Definition of paint. Paint can be defined as a mixture of a vehicle and finely ground white solid particles called pigment. The vehicle is the liquid part of the paint, whereas the pigment can be considered the solid part of the paint that is held in suspension by the vehicle.

Vehicle. The vehicle consists of the film binders (resins or natural oils) that form the film. It holds together the white paint solids and causes the paint film to adhere to the surface. The vehicle also supplies the protective and durable qualities to the film (see Tables 27-1, 27-2, and 27-3).

Pigments. Paints can be colored to almost any shade desired by the addition of color pigments, which are finely ground color solid particles (see Table 27-4).

In order to make the paint dry faster, *driers* are added. There are

Table 27-1.   The Natural Resins

Type	Source	Solvent
Damar	Oriental trees	Alcohol or turpentine
Ester gums	Glycerin, rosin	Alcohol or turpentine
Rosin	Residue from distillation of turpentine oil	Alcohol or turpentine
Shellac	Produced from the lac bug	Alcohol

Table 27-2.   Synthetic Resins

Type	Solvent
Acrylate	Benzene
Alkyd	Water, turpentine
Casein	Water
Chlorinated rubber	Ketone
Epoxy	Ketone, mineral spirits
Latex	Water
Phenolic	Varies with type
Polyvinyl	Water, benzene, and turpentine

Table 27-3.   Natural Oils

Type	Solvent
Raw linseed oil	Turpentine
Fish oils	Alcohol, benzene
Safflower oils	Ether
Soybean oil	Alcohol, ether
Tung oil	Ether

Table 27-4.   Primary Color Pigments

Color	Name	Source or Ingredient
Black	Lamp black	Carbon from burnt tars
Blue	Ultramine blue	Double silicate of sodium and aluminum
Brown	Burnt umber	Iron oxide
Green	Emerald green	Hydrated chromium oxide
Red	Cadmium red	Cadmium sulfide or barium sulfate
Yellow	Yellow ocher	Hydrated iron oxide

also a series of *extender pigments* that can be added which give the film certain improved characteristics (see Tables 27-5 and 27-6).

Hiding power. This refers to covering capacity or the ability to cover a surface so that nothing shows through. All white paints are graded or described on the basis of hiding-power units, which are derived from the hiding power of 1 lb. of white lead: 1 lb. of white lead mixed with linseed oil will completely cover 15 sq. ft. of a surface consisting of alternate stripes of black and white (see Table 27-7 and Figure 27-1).

From this table we can see how titanium compares to lead: a little less than 1/8 lb. of titanium dioxide will completely cover the same 15 sq. ft. of a surface consisting of alternate stripes of black and white.

Today, paints can completely cover a surface with only one coat, can dry within a half hour, and the protective films can be waterproof, fire-resistant, hard and tough enough to become a floor finish, stainproof, antibacterial, and insecticidal. Most paints today come ready-mixed in containers and are color-coordinated. The mixing of paint and the mixing to obtain colors on the job is almost completely nonexistent today.

The most commonly obtainable paints and solvents are shown in Tables 27-8 and 27-9.

Painting

Painting is the word used in specifications to define the application of paint to a surface. The type of paint to be applied must be chosen not only for the decorative effect desired and its protective value but also to fit the type of surface and material to which it is to be applied.

In the construction field there are three methods for applying paints: (1) painting applied on the job, (2) painting applied in a mill, shop, or factory, and (3) prime coats applied in a mill, shop, factory, or on the job.

Painting on the job. This requires that painters not only have brushes, rollers, or spray guns (see Figures 27-2, 27-3, and 27-4) but also a series of accessories necessary to prepare the various surfaces prior to the application of paint:

1. *Thinners* and *solvents* — to thin ready-mix paint and to clean brushes, hands, etc.;
2. *Putty* — to fill holes and cracks;
3. *Caulking compound* — to fill joints between materials;
4. *Wood filler paste* — to fill pores and open grains in wood;
5. *Knot-sealers* — to seal knots and pitch streaks in wood to prevent staining of finish paint;
6. *Paint* and *varnish removers* — to remove old paint and varnish;
7. *Sizing materials* — sealers for plaster, cement, concrete, and masonry materials to seal in the lime, which can destroy or stain the finish paint.

Mill, shop, or factory painting. Shop painting is applied by various methods such as dipping, tumbling, centrifuging, spraying, and knife coating. The types of material and their location in the building control the method of painting to be used.

Application of paints. All painting should be done at temperatures of 60° to 80° F. All surfaces should be clean, dry, prime-coated, sized if required, have cracks filled and rough projections removed. Knots in wood must be sealed and rough grain must be sanded smooth. In masonry, efflorescence must be removed, joints pointed, and excess mortar removed. The areas where painting is to be done must be well-ventilated. Surfaces not to be painted must be protected; and painted surfaces should be protected until the applied paint dries.

Table 27-5. Driers

Type	Qualities
Calcium and cobalt salts	Fast-drying; increases film hardness
Zinc salts	Relatively fast-drying; increases film hardness

Table 27-6. Extender Pigments

Type	Qualities Added to Paint
Barite	Better brushing
Diatomite	Gives flatness and reduces cracking
Kaolin	Strengthens film and reduces settling
Whiting	Adds gloss and increases durability of film

Table 27-7. Hiding Power Units of Various White Paint Solids (pigments)

Type of White Paint Solid (Pigment)	Hiding Power Units
Antimony trioxide	2.33
White lead	1
Zinc oxide	1.48
Titanium dioxide	8.16
Lithopone Zinc sulfide Barium sulfate	1.79

Table 27-8. Common Types of Paint Used in the Construction Field

Type of Paint	Ingredients
Pigmented paints	Ready-mixed in containers — consists of white paint solids, color pigments, extender pigments, vehicle, and driers
Enamels	Ready-mixed in containers — consists of white paint solids, color pigments, a varnish vehicle, driers, and solvent
Varnish	Ready-mixed in containers — consists of a resin, drying oil, drier, and solvent
Lacquers	Ready-mixed in containers — consists of nitrocellulose with resins and drying oils added to improve adhesion and flexibility of film
Shellac	Ready-mixed in containers — consists of refined lac resin and alcohol
Stains	In containers — consists of dry color pigments in a vehicle of water and alcohol or drying oils

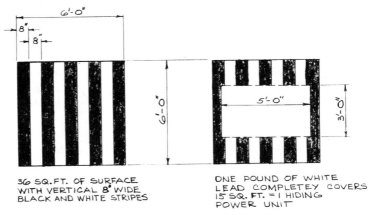

36 SQ. FT. OF SURFACE
WITH VERTICAL 8" WIDE
BLACK AND WHITE STRIPES

ONE POUND OF WHITE
LEAD COMPLETEY COVERS
15 SQ. FT. = 1 HIDING
POWER UNIT

Figure 27-1.   Method of Deriving One Hiding Power Unit

Table 27-9.   Solvents (Thinners)

Type	Type of Vehicle
Acetone	Synthetic resins
Alcohol	Shellac, some drying oils, and some natural resins
Benzene	Drying oils, natural and synthetic resins
Ethyl alcohol	Drying oils, natural and synthetic resins
Ketone	Drying oils, natural and synthetic resins
Mineral spirits	Drying oils, natural and synthetic resins
Water	Water and emulsion-base vehicles

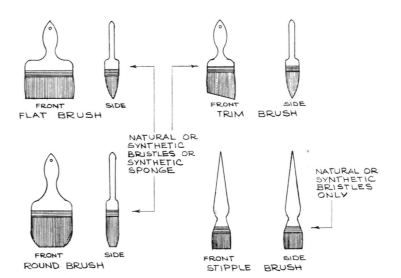

FRONT
FLAT BRUSH   SIDE

FRONT
TRIM BRUSH   SIDE

NATURAL OR
SYNTHETIC
BRISTLES OR
SYNTHETIC
SPONGE

FRONT
ROUND BRUSH   SIDE

FRONT
STIPPLE BRUSH   SIDE

NATURAL OR
SYNTHETIC
BRISTLES
ONLY

Figure 27-2.   Paint Brushes

FLAT ROLLER   TRIM ROLLER

Figure 27-3. Paint Rollers

Figure 27-4. Paint Spraying

REVIEW EXAMINATION

1. Paint is a mixture of what two components?

2. The vehicle is what part of the paint?

3. What can be added to make paint dry faster?

4. One lb. of white lead mixed with linseed oil is the basis of what?

5. Name two methods for applying paint.

6. Name five accessories a painter uses in applying paint.

7. Name three methods by which paint is applied.

8. At what temperature should paint be applied?

9. Name two surface preparations necessary before paint is applied.

ASSIGNMENT

1. Make up a list of materials used in the construction field that require a shop coat of paint.

2. Make a sketch showing the three different methods of applying paint.

SUPPLEMENTARY INFORMATION

Shape of White Paint Solids

The shape and size of the white paint solids (pigments) play an important part in the hiding power and the smoothness, strength, thickness, and resistance to moisture of the paint (see Figure 27-5). Particle size varies in diameter from 0.0001 to 0.0060 millimeters.

Varnish

Varnish is used not only as a clear transparent coating but also as the vehicle for interior pigmented (colored) enamels. In addition, it gives smoothness and leveling characteristics to the surface film. The word "length," when used in relation to varnish, means the ratio of oil to resin, based on the number of gals. of oil to 100 lbs. of resin (see Table 27-10).

Painting Ferrous Metals

All ferrous metal, except stainless steel and special steels (Cor-ten) require special rust-inhibiting paints and must be cleaned by special

TYPES OF PARTICAL SHAPES		
TYPES	SHAPE	EFFECT ON PAINT FILM
ROUNDED	⟲	THICKNESS
NEEDLE-LIKE	 ▭	STRENGTH
PLATE-LIKE	 ⬭	RESISTANCE TO MOISTURE

Figure 27-5. Paint Particle Shapes

Table 27-10.   Lengths of Varnish

Gallons of Oil	Pounds of Resin	Length of Varnish
Less than 20	100	Short-oil
20 to 30	100	Medium-oil
30 or more	100	Long-oil

Table 27-11.   Methods of Cleaning Ferrous Metals

Name of Method	Where Done	Method
Flame	Shop, mill, or factory	Oxyacetylene flame
Pickling	Shop, mill, or factory	Metal submerged in phosphoric or sulfuric acid solution
Sand-blasting	Shop, mill, or factory	Jet of sand blown with compressed air
Wire brushing	Shop, mill, factory, job site	By hand on job and by machine in shop, mill, or factory
Rust remover	Job site	Applied by brush and wiped clean with cloths
Solvent cleaning	Job site	Applied by brush and wiped clean with cloths

Table 27-12.   Rust-Inhibiting Paints

Type	Color
Red lead	Bright orange-red
Blue lead	Gray blue
Zinc chromate	Yellow

Table 27-13.   Dampproofing Paints

Type of Paint	Ingredients
Bituminous Paints	Coal-tar pitch with a solvent
Asphalt paints	Various forms of asphalt with a solvent
Cement paint	Portland cement and lime as the white paint solids, with water as the vehicle and including color pigments, extender pigments, and water repellents

methods (see Tables 27-11 and 27-12).

Prime coating or protective coating is required for both ferrous and nonferrous metal. Ferrous metals (other than stainless steel and special ferrous metals such as Cor-ten) require a prime coat to be applied by the manufacturer for two purposes: to protect the metal until it is installed, and to provide a painted surface on which a finish coat of paint can be applied on the job site.

Nonferrous metals are given a protective removable coating which is taken off when the possibility of damage or staining is no longer present at the job site.

Dampproofing Paints

Dampproofing of foundation walls and exterior surfaces of buildings is considered painting. This is done with special paints, and the painting of interior concrete and masonry is, in many areas, done with cement-type paints (see Table 27-13).

Colors and Mixing Combinations

The primary colors are red, yellow, and blue; secondary colors are green, violet, and orange; and tertiary colors include red-orange, yellow-orange, yellow-green, blue-green, blue-violet, and red-violet. Black and white are classified separately. Table 27-14 shows mixing combinations, wherein the combining of two or more primary colors results in a secondary color, and the mixing of primary and secondary colors results in tertiary colors. Because a prerequisite to passing the contacting examination pertains to the mixing of colors, this information is especially valuable to contractors and painting contractors.

Table 27-14. Colors and Mixing Combinations *

Principal Color	Colors Added to Principal Color	Resulting Color
White	Yellow ochre and red	Buff
Red	Black and yellow	Chestnut
Raw umber	Red and black	Chocolate
Red	Umber and black	Claret
Red	Yellow and black	Copper
White	Vermilion, blue, and yellow	Dove
White	Yellow ochre, red, and black	Drab
White	Yellow and red	Fawn
White	Yellow ochre and vermilion	Flesh
Red	Black, yellow ochre, and white	Freestone
White	Prussian blue and lake	French gray
White lead	Black	Gray
White	Stone ochre and red	Gold
Chrome green	Black and yellow	Green bronze
White	Chrome green	Green pea
White	Chrome yellow	Lemon
White	Yellow ochre, black, and red	Limestone
Yellow	Blue, black, and white	Olive
Yellow	Red	Orange
White	Vermilion	Peach
White	Black and blue	Pearl
White	Vermilion and lake	Pink
Violet	Red and white	Purple
Red	Blue and white	Violet

*The first-named color is the principal ingredient, and the others follow in order of their importance. Exact proportions of each color must be determined by experiment with a small quantity. It is best to have the principal ingredient thick and the others thin when mixing.

28

Flooring and Installation of Flooring

INTRODUCTION

The type of flooring and the method of installation must be selected for durability, hardness, resilience, color, comfort, and maintenance. This unit will cover the various types of flooring such as seamless, carpet, wood, tile, terrazzo, resilient, etc., and their installation.

TECHNICAL INFORMATION

Flooring

Definition. Flooring may be defined as the finished surface upon which people walk and furniture is placed, and upon which materials and equipment are carried, rolled, moved or stored.

Criteria for selection. When selecting various types of flooring materials for buildings, there are specific requirements that should always be considered:

1. All types of flooring require that the surface upon which they are applied and their method of installation must be suited to correctly receive the selected flooring. For example, wood floors require that water, moisture, and dampness must be prevented from getting into and under the wood flooring. Figure 28-1 shows a method of accomplishing this by using a waterproof adhesive and installing a moisture and vapor barrier over porous fill before laying the floor.

2. Expansion and contraction should be considered not only for the flooring material itself but also for the materials and surfaces upon which the flooring is to be installed. For example; terrazzo must be subdivided with divider strips and installed on a sand bed to take care of the expansion and contraction of the terrazzo and of the material and surface upon which it rests. This prevents the terrazzo from cracking (see Figure 28-2).

3. Different levels will be required in the subfloor surfaces to meet the various depths required by different floor materials for installation. For example, quarry tile, when set in mortar, requires a depth of 1-1/4 to 2-1/2 in., whereas 1/8-in. thick vinyl asbestos tile requires none; therefore, the subsurface for the quarry tile must be depressed between 1-1/8 to 2-3/8 in. below the surface required for vinyl asbestos tile (see Figure 28-3).

4. The treatment of the joints where different types of flooring meet should be carefully detailed. For example, vinyl tile meets ceramic tile flooring as shown in Figure 28-4.

5. The final finish treatments (maintenance) such as waxing, polishing, sealing, etc. must be correct for the flooring. For example, the wax used for wood floors cannot be used on asphalt tile because it has a petroleum base and petroleum products dissolve apshalt tile.

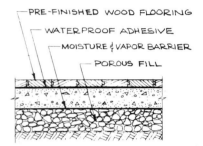

Figure 28-1. Wood Flooring

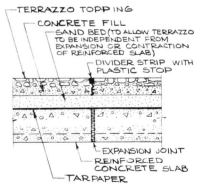

Figure 28-2. Terrazzo Flooring

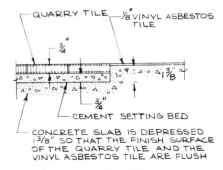

Figure 28-3. Tile Flooring

231

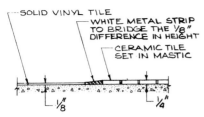

Figure 28-4. Tile Flooring (cont.)

Types of flooring. Flooring materials are summarized in Table 28-1 by their qualities and where they are generally used in building construction.

Methods of Installation

Carpet. The development of synthetic fibers has changed the entire concept of carpeting. Formerly, almost all carpets were manufactured from wool and other natural fibers and were made to specific sizes. Today, a single roll of 15'-0"-wide carpet can be 90'-0" long and longer, and when joined, no seam is visible.

The fibers generally used for manufacturing carpet are polyester, acrylic, nylon, and wool. The carpets with synthetic fibers can be installed both indoors and outdoors. They are resistant to color fading, mildew, rot, moth, bacteria, insects, stain, and flame. They are easily maintained and can be installed in areas of from light to heavy wear. Depending on usage, their durability varies from five to fifteen years.

All types of carpet must have an underlayer, either as an integral part of the carpet or as a separate material. In general there are three methods of installing carpet: stretching and tacking, adhesive, and double-faced adhesive tape (see Figure 28-5).

Ceramic and quarry tile. This type of flooring is durable, hard, and easily maintained and is generally used in areas where water and moisture are present. There are two methods of installing this type of flooring: by the adhesive method and by the cement mortar bed method (see Figure 28-6).

The surface upon which this type of flooring is installed should be rigid so that no bending can occur, thus avoiding cracks that may develop in the joints. Where water and moisture are present, the top of the subsurface should be waterproofed.

Concrete. When concrete is to be used as the finish floor surface, it must be treated with hardeners, either mixed with the concrete or applied during curing or after the concrete is cured. This is necessary because concrete, as it wears, gives off a fine dust, — that is, "dusting" occurs — which hardeners eliminate.

The concrete surface may be colored, wear-resistant, nonsparking, and nonslip. These various surfaces are achieved by the following means:

Table 28-1. Types of Flooring

Material	General Area of Use	Qualities
Carpet	Where resilience, quietness, and limited durability are required. Some types will stain.	Limited durability, fairly easily maintained
Ceramic and quarry tile	Where water and moisture are present and sanitary conditions are required	Durable, hard, and easily maintained
Concrete and concrete with surface treatments	General utility areas and storage, and where there are manufacturing and industrial processes	Hard, durable, requires surface treatment to harden and eliminate dust and will stain
Resilient: asphalt, cork, rubber, vinyl, vinyl asbestos tiles, asphalt, and linoleum	Where resilience is required and where there will be slight to moderately heavy wear	From very resilient to slightly resilient (depending on type) and from poor to easily maintained (depending on type)
Paint: oil-based	General utility areas and where there will be slight wear	Limited durability, difficult to maintain, requires periodic repainting
Paint: synthetic resin-based (epoxy acrylic, etc.)	General utilities areas where there will be slight to heavy wear	Requires refinishing periodically
Seamless-epoxy, polyester, oxychloric, and other types of synthetic resins	Areas where there will be slight to very heavy wear	Durable, resilient to hard, easily maintained
Steel	In areas where there are manufacturing and industrial processes and also for supporting heavy equipment	Very strong, durable, but requires periodic repainting
Stone and brick	Where there will be very heavy wear also where water and moisture are present	Durable, hard, good maintenance but will stain
Terrazzo: all types	Where there will be very heavy wear	Durable, hard, easily maintained but will stain
Wood: all types	Where there will be slight or very heavy wear (depending on type)	Fairly to very durable, relatively easily maintained, but will stain

1. *Metallic oxide colors* can be mixed with the concrete or applied to the surface when it is floated, and troweled to give color.

2. *Metallic aggregates* can be applied to the surface when it is floated and troweled to give a wear-resistant surface.

3. *Metallic iron aggregates* (to give a nonspark flooring) are applied in the same manner as metallic aggregates, and the entire surface treatment is electrically grounded. Nonsparking finishes are used for areas where highly explosive or flammable materials are to be stored or used.

4. *Abrasive aggregates* are applied in the same manner as metallic aggregates to make a very nonslip surface (see Figure 28-7).

Painted floor surface. Wood and concrete have been painted for many years with oil-based paints as a finish surface flooring for cleanliness and maintenance in areas where materials are stored and in utility areas. Paint should be used only in areas where there will be very slight wear. It requires periodic repainting.

Synthetic resin based paints can be used as a finish surface flooring for areas where heavy wear will be present. These paints are hard, durable, and nonslip; they are available clear and colored, and can be applied to wood, concrete, and metal.

This type of surface flooring consists generally of two materials in separate containers — the synthetic resin and the activator. When thoroughly mixed, the end product is applied by brush, roller, or spraying and requires extensive ventilation during application because most of these paints are highly volatile and flammable. Great care should be taken that they be installed immediately once mixed because of their quick drying (known as pot-life). The usual pot-life is around 45 minutes — after this they cannot be used and must be confiscated and thrown away.

Seamless flooring. This type of flooring has been installed for many years, using oxychloric cement as the cementitious material with wood sawdust, sand, and other ma-

terials mixed and applied 1/2 to 3/8-in. thick and troweled level. The finished flooring is resilient,

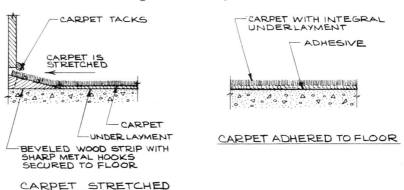

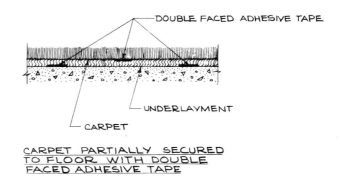

Figure 28-5. Installation of Carpet

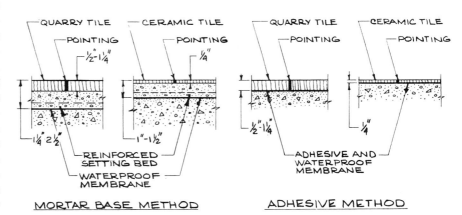

Figure 28-6. Installation of Quarry and Ceramic Tile

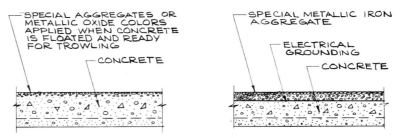

Figure 28-7. Application of Special Surface Treatments for Concrete

noncombustible, nonslip, and easy to maintain. It will endure light to heavy wear.

The development of synthetic

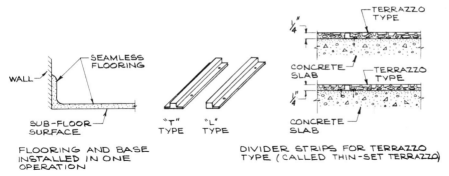

Figure 28-8. Seamless Flooring

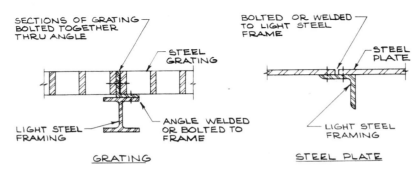

Figure 28-9. Steel Flooring

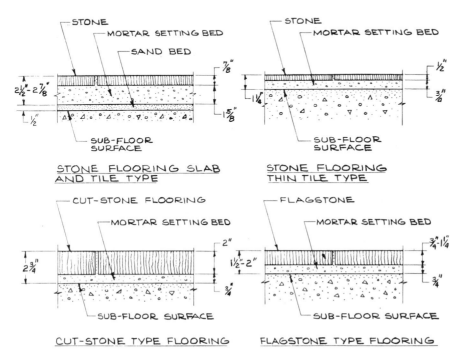

Figure 28-10. Stone Flooring

resins has resulted in a wide variety of seamless floors which can be very hard to quite resilient, also nonslip and spark-resistant. They are available in a wide variety of colors and can be used in areas where light to heavy wear will be present.

The cementitious synthetic resins generally used are epoxy, vinyl, urethane, acrylic, and neoprene. They are combined with colored plastic flakes, cork, ceramic-coated stone, and marble chips. The thickness can vary from 1/16 to 3/8 in., depending on type used. When synthetic terrazzo-type seamless floors are installed, divider strips are sometimes used but are not necessary.

One great advantage of seamless floors is that they can continue up the wall to base height, thus eliminating the entire problem of having to install the base after the floor has been laid (see Figure 28-8).

Steel. Steel flooring is manufactured in two types, grating and steel plate. All steel flooring must be supported by light steel framing (see Figure 28-9). Steel plate is available with various surface patterns and textures to make it nonslip.

Stone and brick flooring. All types of stone and brick for flooring are almost always set with a cement mortar bed. The stones used for flooring are marble, granite, limestone, bluestone, and slate. Of these stones, granite, bluestone, and slate are installed both indoors and outdoors, whereas marble and limestone are installed indoors (see Figure 28-10).

Face brick and paving brick are both used for flooring. Face brick is generally manufactured with holes; therefore, when used for flooring, only the 2-1/4-in. face can be exposed as the finish surface. Flooring brick is a dense, hard-burned, impervious type of brick that may be obtained acid-resistant, generally 8-1/2 in. long × 4 in. wide × 2-1/2,

3, and 4-1/2 in. deep (see Figure 28-11).

Resilient flooring. The various types of resilient flooring include asphalt, cork, rubber, vinyl, vinyl asbestos, and linoleum. They are manufactured in a wide range of colors, designs, patterns, and textures.

All the various resilient floorings are manufactured in two forms, either as tiles, or in rolls, or both. The tile shapes are generally squares 9 × 9 in. and 12 × 12 in., in thicknesses of 1/16, 3/32, and 1/8 in. The rolls are generally 6'-0" wide and vary in length from 30'-0" to 50'-0". Their thicknesses are 1/16, 3/32, and 1/8 in. In general the 1/8-in.-thick tile or roll is for areas where heavy wear will be present. These floorings are durable, resilient (varying from quite resilient to almost nonresilient) and easily maintained. They can be used in areas of slight to heavy wear.

The manufacturers of resilient-type flooring make bases, feature strips, and special designs for inlaying. Custom designs are also possible. As an example, on a playroom floor, checkerboard or various game markings can be installed.

When selecting a resilient type of flooring, the surface upon which it is to be applied should be smooth, have no cracks, and be protected from moisture and water (see Figure 28-12).

Terrazzo. There are two types of terrazzo flooring, one in which portland cement is the cementitious material and one in which a synthetic-type resin is the cementitious material. In both types, marble chips are added and, when cured, the surface is ground to a fine, smooth finish. Terrazzo flooring is very hard, durable, easy to maintain, and able to withstand very rough wear. The various colors are obtained by combining colored marble chips and also by coloring the cementitious material. Terrazzo is installed either at the site or as prefabricated units (see Figure 28-13).

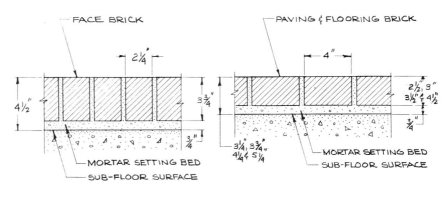

FACE BRICK FLOORING   PAVING OR FLOORING BRICK

Figure 28-11. Brick Flooring

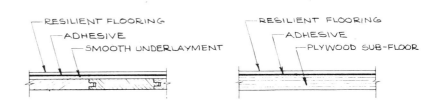

ON T&G SUB-FLOORING   ON PLYWOOD SUB-FLOORING

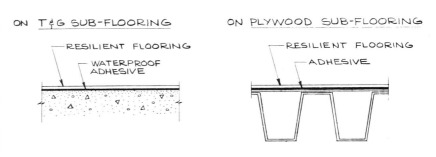

ON CONCRETE SLAB   ON STEEL DECKING

Figure 28-12. Resilient Flooring

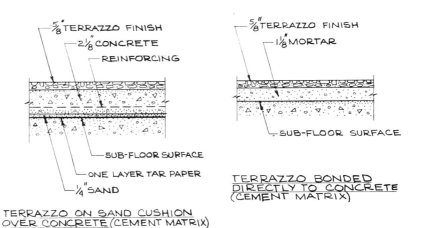

TERRAZZO ON SAND CUSHION OVER CONCRETE (CEMENT MATRIX)

TERRAZZO BONDED DIRECTLY TO CONCRETE (CEMENT MATRIX)

Figure 28-13. Terrazzo Flooring

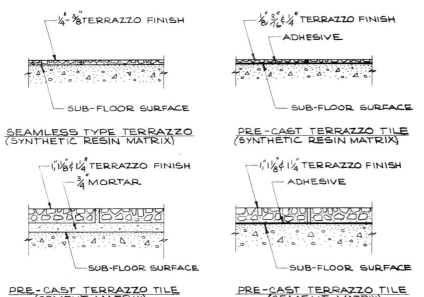

Figure 28-13. Terrazzo Flooring (Cont.)

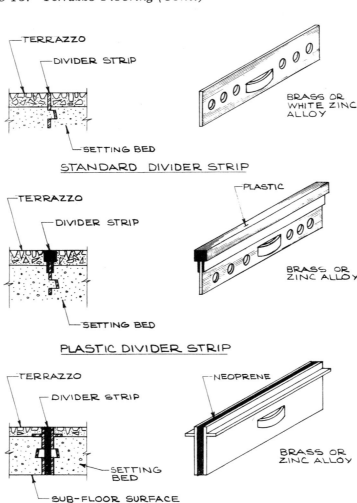

Figure 28-14. Terrazzo Dividing Strips

All on-site cement-matrix terrazzo requires divider strips to stop any cracking of the finish surface (due to expansion and contraction) and any deflection of the supporting subflooring (see Figure 28-14). With the seamless type, divider strips are installed only as a precaution against cracking. With the precast type, they are installed only when there is great length and changes in direction occur.

Wood flooring. The various evergreen and decidious woods used for flooring must be close-grained, hard, and wear-resistant. All wood flooring must meet the rigid rules and regulations for grading, moisture content, and kiln-drying that have been set by the Wood Flooring Manufacturers Associations. The woods generally used are oak, maple, fir, and spruce. There are three types of wood flooring: strip, thin square blocks, and solid end-grain blocks (see Figure 28-15). Strip flooring is manufactured unfinished (must be sanded and finished on job site), and prefinished (no job sanding or finishing required).

Decidious strip flooring is manufactured quarter-sawed in thicknesses of 5/16, 1/2, 3/8, 25/32, 17/16, and 35/32-in., and widths of 1-1/2, 2, 2-1/4, 2-1/2, 3, 3-1/4, and 3-1/2 in.

The grading is controlled by the lengths. For example, *maple, first grade*, shall be 2'-0" and up as stock will produce; not over 30% of total footage shall be in bundles under 4'-0" in length. The grades in general are *first*, *second*, and *third*, but certain types have other grades depending on the species of wood.

Softwood strip flooring is manufactured plain-sawed, in thicknesses of 3/8, 1/2, 5/8, 1, 1-1/4, 1-1/2, 2, 2-1/2, 3, and 4 in., and widths of 2, 3, 4, 5, 6, 8, 10 and 12 in., bundled in various lengths and graded *select* and *common*. All strip flooring is bundled and delivered by the board foot.

Thin types of wood block flooring are prefabricated either from solid wood or veneers (similar to plywood). They are manufactured prefinished in different patterns and generally 9 X 9-in. and 12 X 12-in. squares.

Solid end-grain wood block flooring is installed in areas where very heavy wear will be present, such as in factories and in school shops. (see Figure 28-16).

The surface upon which all types of wood flooring is to be installed must be dry and waterproof so that no moisture or water can be absorbed by the wood (see Figure 28-17).

Wood flooring should be selected: by type of wood, as either quarter-sawed or plain-sawed, by the type of wear that it will be subjected to when installed, by color, by grain, and by the type of surface upon which it is to be installed.

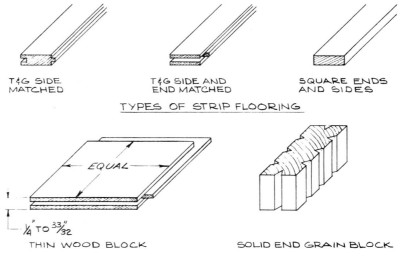

Figure 28-15. Types of Wood Flooring

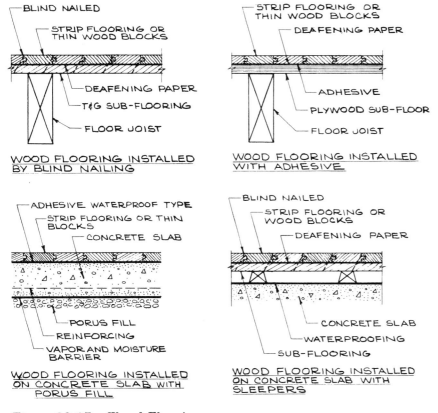

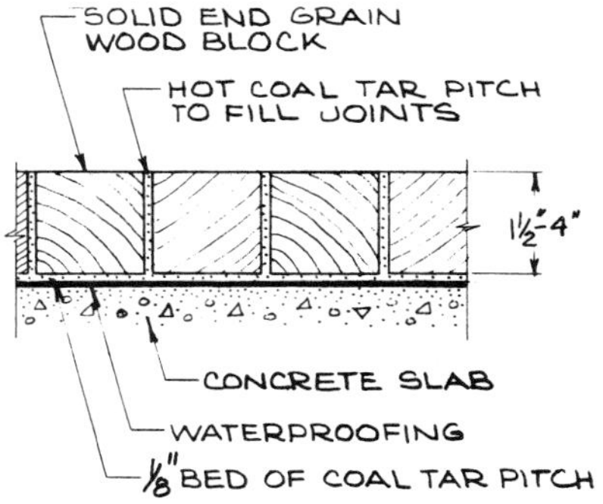

Figure 28-16. End Grain Solid Wood Block Flooring

Figure 28-17. Wood Flooring

237

1. When selecting various types of flooring, are the following statements correct? (Answer yes or no.)

(a) Expansion and contraction are important.

(b) The method of treating the joints between different types of flooring is important.

(c) The surface upon which various types of flooring are to be installed is unimportant.

2. What are the three methods of installing carpet?

3. Can carpet be installed outdoors?

4. What are the two methods for installing ceramic tile and quarry tile?

5. What must be done to concrete when it will be used as a finished floor?

6. Concrete, when used as a finished floor, can be (answer yes or no): (a) colored, (b) made resilient, (c) made nonslip, (d) made nonspark, (e) made acoustic.

7. When oil-based paint is used as a finish flooring surface can it be installed in areas where heavy wear will be present?

8. What kind of wear will be possible when synthetic resin paint is used?

9. What resins are used as cementitious material in seamless flooring?

10. Steel flooring is manufactured in what two types?

11. Stone and brick flooring are installed with what setting bed?

12. Name four different types of resilient flooring.

13. Terrazzo consists of either a portland cement or a synthetic resin cementitious material and what other material?

14. To avoid cracking, what is used with terrazzo flooring?

15. Name the three types of wood flooring.

ASSIGNMENT

1. Make a freehand sketch, in scale, showing the specific information and requirements that should be considered when installing quarry and ceramic-type flooring.

2. Make a freehand sketch, in scale, showing the types of wood flooring.

SUPPLEMENTARY INFORMATION

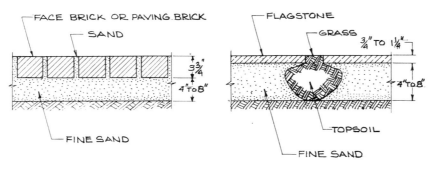

Figure 28-18.   Brick or Flagstone Installed in Sand—Outdoors

Brick and flagstones. When used as flooring on the exterior, brick and flagstones are installed either with mortar as previously stated or in sand (see Figure 28-18).

Design precautions. It is of major importance when various types of flooring are selected for a building that the supporting structure be considered because of the various levels or depressions that may be required for each type of flooring material (see Figure 28-19).

Wood flooring in large areas. When selecting wood flooring for large areas such as gymnasiums or large manufacturing areas, there are special methods of installation, as shown in Figure 28-20.

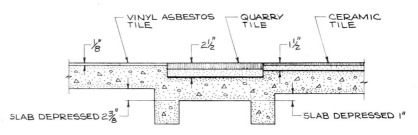

Figure 28-19. Effects of Various Types of Flooring on Supporting Structure

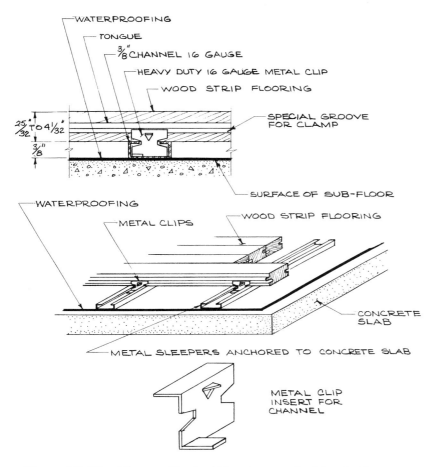

Figure 28-20. Heavy Single Flooring with Metal Sleepers Such As Are Used on Gymnasium Floors

29

Plaster, Plastering, and Surface Treatments

INTRODUCTION

This unit describes plaster as a material, its various types and uses, its application, and the products manufactured from it. In building construction, plaster products represent an increasing use of this material.

The various types of special surfacing materials used on the exterior and interior, other than paint. are also described, because they represent a recent technological development that is finding wide acceptance in the building industry.

TECHNICAL INFORMATION

Plaster

Plaster consists of a mixture of a cementitious material, aggregate, water, admixtures for controlling workability and setting time, and hair and fibers. There are four types of cementitious materials used to make plaster: gypsum, Keene's cement, portland cement, and lime. The aggregates generally used are sand, perlite, vermiculite, and wood fibers. Plastering materials are manufactured either premixed with aggregate and additive in bags or

Table 29-1. Plaster Ingredients

Ingredients	Type and Form	Use
Cement	Portland or portland pozzolana	Exterior and interior walls and ceilings
Fiber	Glass, hemp, sisal, or jute Clean, long, and free from tannic acid	Reinforcing plaster
Gypsum	Plaster of paris	Making gypsum plasters
	Gypsum plaster	2- and 3-coat interior
	High-strength gypsum plaster	Plasterboard, decking, block, 2- and 3-coat interior plastering
	Keene's cement	Finish for 2- and 3-coat gypsum plaster, and Keene's cement plaster
	Fibered gypsum plaster	Scratch coat
	Prepared gypsum plaster	2- and 3-coat interior
	Fire-resistant gypsum plaster	Fireproofing steel, etc.
	Acoustical plaster	Acoustical treatment for walls and ceilings
	Bonding plaster	Finish for concrete walls and ceilings
	Sand-float finish plaster	For finish of 2- and 3-coat plastering
Hair	Goat or cattle 1/2 to 2″ long. Clean and free from rust, knots, and balls	Reinforcing plaster
Lime	Hydrated lime Lime putty	Gives workability and plasticity to plasters
Sand	White or light gray fine sand, 100% passing through a No. 4 sieve	Aggregate for plasters
Water	Clean, fresh water containing no salt, sulfur, or other harmful substances	Makes cementitious materials and aggregate workable, and also combines chemically with gypsum and lime
Lightweight aggregate	Perlite or vermiculite	Imparts fire resistance to plasters

241

barrels, so that only water is added at the site; or without additives in bags or barrels, with other materials separately packaged so that the mixing is done on the site. The ingredients for making plaster are shown in Table 29-1.

It should be noted that gypsum plaster is used not only as a plastering material but also to make decking, plaster board, partitions, blocks, and poured decking.

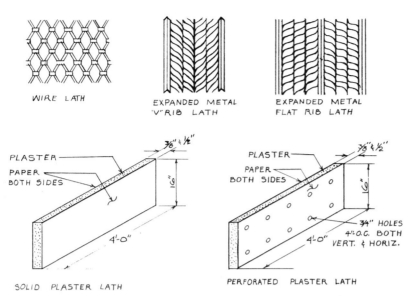

Figure 29-1.   Lath for Plastering

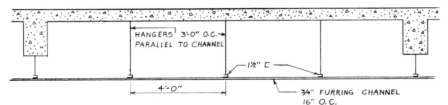

Figure 29-2. Installation of Furring Channels to Receive Wire Lath and Plaster Ceiling

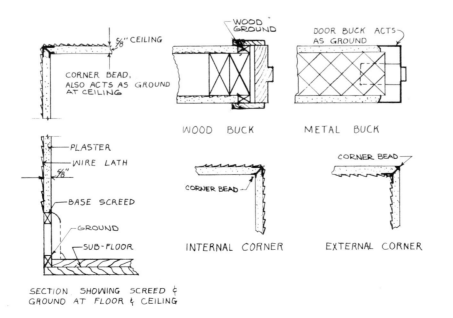

Figure 29-3.   Application of Plaster

Plastering

The word "plastering" refers to the application of the various types of plaster to a surface in layers to give a smooth or textured hard surface finish which can then receive paint, wallpaper, ceramic tile, plastic laminates, textiles, etc., as final finish.

For exterior plastering, portland-type cements are used; where water and moisture are present on the interior, Keene's cement-type plaster is used; and for interior plastering, gypsum plaster is generally used.

The plaster can be applied to various types of masonry walls such as brick, structural clay tile, gypsum block, etc., either directly or onto wire lath, plaster-board lath, or other material that will give sufficient keying or bonding for the plaster. The thickness of plaster is usually 5/8 to 7/8 in. on masonry and wire lath and 1/2 in. on plaster lathing board and gypsum block.

Plaster is applied either by the two-coat or three-coat method. The first coat is called the *scratch coat*, the second coat is called the *brown coat*, and the finish coat is called *hard finish*. In two-coat work the scratch and brown coats are combined and are called the base coat, with the finish coat the same as that used for three-coat plastering (see Table 29-2).

Table 29-2.   The Thickness of Plastering Coats

Three Coats	Two Coats
Scratch coat 1/2″	Base coat 1/2″
Brown coat 1/4″	Hard finish 1/8″
Hard finish 1/8″	

Lath, either metal or plasterboard, is used to span the open spaces between structural framing to form a surface to which plaster can be applied. All lath must be of a type to which the plaster can bond (see Figure 29-1).

For ceilings where structural members are widely spaced and for hung ceilings, a system of small channels must be installed to which the lath can be applied (see Figure 29-2).

Before plaster is installed, grounds and screeds are employed, set in such a way that the plaster will have a level and plumb surface (see Figure 29-3). The various accessories used in plastering to take care of corners, expansion, reinforcing, changes in materials, etc., are manufactured for both two-coat and three-coat plastering (see Figure 29-4).

Plaster Products

Gypsum boards. These consist of a core of gypsum plaster containing fibers or hair sandwiched between two sheets of heavy paper, usually with sized paper on one side and unsized paper on the other. The core may include perlite or vermiculite and glass or mineral fibers, all of which increase fire resistance.

Gypsum boards, more commonly known as *sheetrock*, are manufactured in various types: regular, fire-rated (type X), backer board, insulating, vinyl-coated, sheathing, decorative, and laminated for non-load-bearing partitions.

Regular sheetrock has one side with sized paper, the other unsized paper, and is available in thicknesses of 3/8, 1/2, and 5/8 in., 4'-0" wide and in lengths 6'-0" to 14'-0" in 1-ft. increments. In residential and other wood-frame construction work, gypsum boards are nailed directly to the wood frame as a surfacing material for walls and ceilings to receive paper, paint, and textiles. The regular type is also used extensively in larger construction on metal partitions; it is made to adhere to rigid foam insulation on masonry walls and to prefabricated laminated sheetrock partitions (see Figures 29-5 and 29-6).

Fire-rated (type X) gypsum consists of sheetrock with one side of sized paper and the other of unsized paper; it is available in thicknesses of 1/2 in. with a 45-min. fire rating and 5/8 in. with a 1-hr. fire rating; 4'-0" wide and in lengths 6'-0" to 14'-0" in 1-ft. increments. This type is used in areas that require 45-min. or 1-hr. fire resistance.

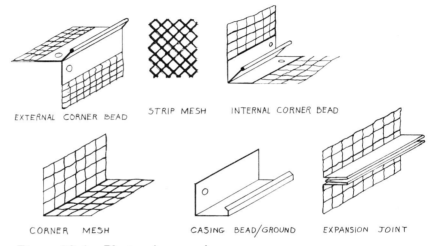

Figure 29-4. *Plaster Accessories*

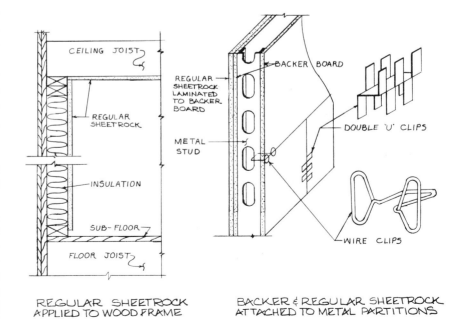

Figure 29-5. *Details of Application of Sheetrock*

Backer board consists of sheetrock with unsized paper on both sides, available in thicknesses of 3/8, 1/2, and 5/8 in., 2'-0" wide, and 8'-0" long. It is used primarily as a backing material to which other finish materials are applied — for example, as backing for acoustical tile applied with an adhesive.

Insulating sheetrock consists of any of the various types with aluminum foil laminated to one side. The aluminum foil acts as a vapor barrier and reflective-type insulation.

Vinyl-coated sheetrock has one side treated with a waterproof vinyl coating; it is available 3/8 and 1/2 in. thick, 4'-0" wide, and 8'-0" long. Its major use is in areas where moisture and vapor are present; it also is used as backing for adhesive-type installation of ceramic, metal, or plastic tile.

Sheetrock *sheathing* consists of a core treated with asphalt and the outside paper surface treated for water resistance; it is 1/2 in. thick, 2'-0" wide, and 8'-0" long. It is used for exterior sheathing for frame construction.

Decorative sheetrock has one side treated with a prefinished surface treatment (decorative paper, wood grains, etc.).

Laminated sheetrock partitions consist of two, three, or four sheets of sheetrock laminated together to manufacture prefabricated partitions (see Figure 29-6). There are various types of partition systems, both permanent and removable, using sheetrock.

Edges and finishes. The various types of sheetrock have different types of edge treatment (see Figure 29-7). The finish treatment of joints and nail indentations consists of what is known as *spackling* (see Figure 29-8).

Decking. Gypsum plaster decking is manufactured 2 in. thick, 15 in. wide, and 10'-0" long and generally has a metal tongue-and-groove frame at its edges. The maximum span between supports is 7'-0". Gypsum plaster decking can be cantilevered 2'-6" in its 10'-0" dimension and only 6 in. in its 15-in. dimension (see Figure 29-9).

Poured roofing. Poured gypsum roofing is lightweight, fire-resistant,

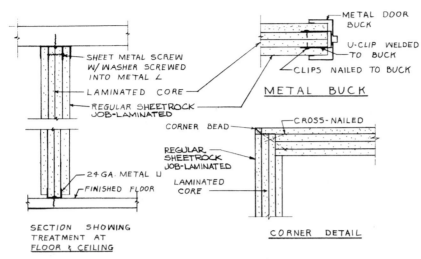

Figure 29-6. Details of Prefabricated Partition

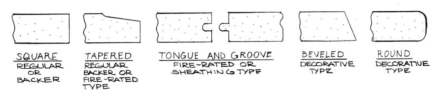

Figure 29-7. Edge Treatments for the Various Types of Sheetrock

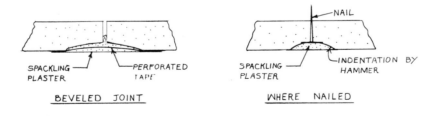

WHEN SPACKLING PLASTER IS THOROUGHLY DRY IT IS SANDED SMOOTH

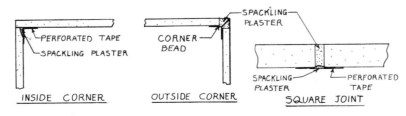

Figure 29-8. Sheetrock Edge and Corner Treatments

has the advantage of setting in a half-hour, and develops enough heat to be installed in freezing weather. It is generally used for roof construction because of its setting time and light weight and also because it can easily be mixed at ground level and pumped directly to the roof. Its one disadvantage is that, if wet, it can deform; therefore, the finished roofing material should be applied as fast as possible. A poured gypsum roofing job, if well-organized, can be completely installed with finish roofing in one day (see Figure 29-10).

The wire lath generally used to reinforce poured gypsum roofs is No. 12 gauge wire, 4-in. OC lengthwise, and No. 12 gauge wire, 8-in. OC crosswise. The materials used for the permanent form depend upon the type of interior finish required. An example would be 1-in. insulating form board with acoustical tile applied with adhesive.

Surfacings

"Surfacings" is the word adopted by the building industry for special types of surfacing finishes. Many of these "surfacings" consist of the same synthetic resins, adhesives, and synthetics used in paints and coatings; but when used in combination with other materials, they form new types of surface finishes. For example, epoxies mixed with cement and fine sand form a paste which can be troweled to give a waterproof, durable, colored, textured finish surface.

Surfacings can be factory- or shop-applied and job-site applied. They are available in a wide variety of finishes, textures, and colors; they can be waterproof, are easily cleanable, dustproof, and fire-retardant; they can serve as breathers (to allow moisture vapor from back-up material to be dissipated) and be resistant to shock and abrasion.

Surfacings can be brushed, sprayed, troweled, and rolled; the exposed aggregates, when used, can be seeded (embedded) or blown on (see Figure 29-11). Generally, all types of surfacings become the permanent surface in 24 hours.

One can anticipate the development of new types of surfacings and special coatings, because much research is being done by manufacturers and there seems to be a general increase in the use of such surfacings.

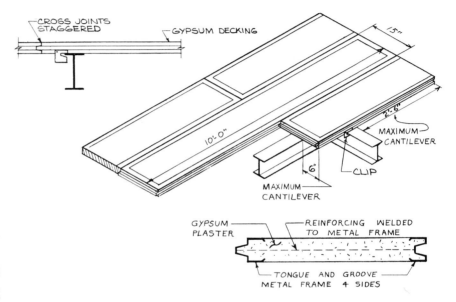

Figure 29-9.   Gypsum Plaster Decking

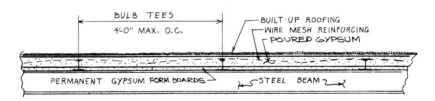

Figure 29-10.   Poured Gypsum Roof Deck

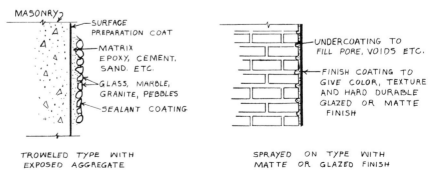

Figure 29-11.   Application of Surfacings

1. Which of the following materials are manufactured from gypsum plaster? (a) plasterboard, (b) asbestos cement board, (c) decking, (d) plaster for the exterior, and (e) partitions.

2. What type of plaster is used for the following areas? (a) exterior, (b) interior, (c) where water and moisture are present.

3. What is the thickness of plaster on masonry and wire lath?

4. Name two types of lath used for plastering.

5. Name two types of accessories used in plastering.

6. Fire-rated (type X) 5/8-in. gypsum plasterboard has what fire rating?

7. What is the finish treatment of joints and nail indentations in gypsum plasterboard called?

8. Gypsum plaster decking is 2 in. thick, 15 in. wide, and 10'-0" long. What is the maximum span between supports?

9. Can poured gypsum plaster roofing be installed in freezing weather?

10. Name four types of gypsum plasterboards.

11. How are gypsum plasterboard partitions manufactured?

ASSIGNMENT

1. Make a freehand sketch to scale, showing details of installing prefabricated gypsum plaster partitions.

2. Make a freehand sketch to scale, showing details for the installation of a poured gypsum plaster roof.

SUPPLEMENTARY INFORMATION

Fireproofing coatings. Such coatings are a special type of surfacing applied like paint; they are only 3/16 in. thick but will give a 2-hour fire rating. These coatings, when exposed to fire and heat, expand from 3/16 to 3 in. of a foam-type insulation. This reaction is known as *intumescence*.

Bonding qualities of surfacings and coatings. The bonding qualities of these new surfacings and coatings are such that they will adhere to almost any material permanently, including wood, concrete, brick, metal, and plaster.

30

Insulation and Acoustical Materials

INTRODUCTION

Many of the insulating and acoustical materials used in the building industry are manufactured from the same raw materials. For example, fiberglass insulation and fiberglass acoustical tile are basically of the same raw material. This unit will deal with broad basic types of insulating and acoustical materials used in the building industry, and their installation and application. Included are flexible, loose, and rigid types of insulation and the various acoustical materials such as tiles, perforated materials, plaster, baffles, and site-assembled materials.

TECHNICAL INFORMATION

Insulation

Most materials offer some resistance to the passage of heat. Those that offer high resistance to the flow of heat are called thermal insulation or simply "insulation." Insulation is needed in all buildings that are exposed to the low temperatures of a winter season or to the high temperatures of a summer season. The more severe the heat or cold, the greater the need for insulation. Insulation primarily adds to the comfort of the individual, and saves both heating and air-conditioning costs.

Rating of insulation. Standard methods of testing insulating materials have been established that give a comparison of the rate at which heat will pass through a given thickness. The rate of heat flow is represented by numbers such as 0.18 and 0.24. These numbers are known as *heat loss coefficients*, or *U-factors*, which, if multiplied by the area of any surface (wall, floor, or ceiling) and by the temperature difference between the inside and outside, will give the heat loss or heat gain in BTUs (British Thermal Units). The formula may be stated as follows:

$$BTUs = U \times TD \times A$$

in which

BTU = the required heat loss or heat gain

U = heat loss coefficient (assigned to materials)

TD = temperature difference between the outside and the inside of the room. Interior design temperature is usually taken at $70°$ F.

A = area of the wall, ceiling, or floor where a known temperature difference exists and where a heat loss occurs

Table 30-1. *Sizes, Covering Capacity and U-Factors of Insulating Materials*

Type	Thickness	Width	Length	Pcs. per Pkg.	Sq. Ft. per Pkg.	U-factor
Blanket quilts	1-1/2"	48"	50'	1	200	0.24
	2"	15"	80'	1	100	0.20
	2"	15"	60'	1	75	
		23"	52'	1	100	0.20
	3"	15"	40'	1	50	
		23"	39'	1	75	0.18
Batt	1-1/2"	15"	96"	8	80	0.24
	2"	15"	24"	24	60	
		23"	48"	8	60-1/2	0.20
	3"	15"	24"	16	40	
		15"	48"	8	40	0.18
Loose	Bag covers approx. 25 sq. ft. of 4" thickness					0.26
Urethane foam (panel)	1"	48"	96"	1	32	0.16
Spray-type insulation	1"					0.26
Polystyrene foam	1"	48"	96"	1	32	0.20
Perlite Vermiculite	Bag fills 50 sq. ft. of 1" space					0.24
Foam glass	1"	18"	24"	1	3	0.35
Insulating board	1"	24"	48"	1	8	0.36

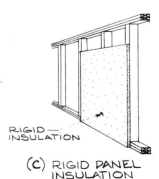

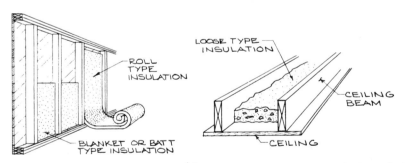

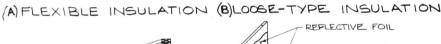

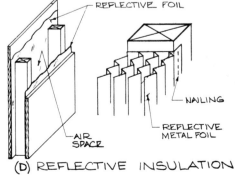

Figure 30-1. *Types of Insulation*

Example.

A room has one exterior wall which is 8'-0" high × 15'-0" long, or 120 sq. ft. in area. The outside design temperature for a specified locality is, say 5° F. above 0° and the inside temperature is 70° F. The difference then is 65° F. The U-factor for the wall is assumed as 0.18. Find the BTU heat loss.

$$BTU = U \times TD \times A$$
$$BTU = 0.18 \times 65 \times 120$$
$$BTU = 1404 \text{ (the required heat loss)}$$

From the above example, it can readily be seen that the lower the U-factors, the better the materials' insulating qualities. Table 30-1 gives the U-factors for various insulating materials, together with the thickness, width, lengths, and covering capacities.

Types of Insulation

Flexible insulation is usually in the form of "blankets" or "quilts," either of which may be in rolls several hundred ft. long or "batts" up to 48 in. long. Because flexible insulation should be built into the hollow spaces in walls or ceilings, the blankets or batts are sized in width to allow them to be fitted into the space between structural framing members. This form of insulation is available in thicknesses from 1/2 to 3-5/8 in. or thicker. One surface of the blanket or batt is usually covered by asphalt-impregnated paper, which serves also as a vapor barrier (see Figure 30-1A). Metal foil is an excellent vapor barrier and is frequently used with blanket or batt insulation for that purpose.

Loose (fill-type) insulation is usually delivered in bales or bags. It can be poured into place or packed by hand into the small spaces around window frames or chimneys. For example, this type of insulation is used to build up any thickness of insulation desired on horizontal surfaces (see Figure 30-1B).

Both flexible and fill-type insulation are frequently made of mineral wool, which is a fluffy or fibrous material made from slag, glass, or rock. Flexible forms of insulation are also made from processed vegetable or animal fibers such as wood, cotton, marine plants, and cattle hair.

Rigid insulation is usually made from some form of wood or cane fiber, glass, and synthetics. It is available in a wide range of sizes, the most common size being 4 ft. wide and 8 ft. long. These insulating boards are usually from 1/2 to 1 in. thick and thicker. They may be smooth or rough, depending on whether they are to be left exposed or to be concealed as part of the wall, foundation, or roof construction. They may be used as plaster base (lath); as base for finishing materials; as sheathing, replacing wood; as a room finish in the form of panels or planks; as roof insulation; and as perimeter insulation for foundations (see Figures 30-1C and 30-2).

Foam insulation is a relatively new insulation material and because of its low U-factor, inertness, fire resistance, and other advantageous properties, it is finding very wide use in the building industry and is replacing many of the existing materials.

The synthetics, urethane and polystyrene, are the most widely used for manufacturing foam-type insulation. They consist of two liquids which, when mixed, form a closed, cellular, rigid mass thirty or forty times its original volume, in which 95% of the total volume is the millions of tiny sealed cells. There are four types of foam insulation that involve different installation methods.

1. *Prefabricated rigid board.* This is manufactured in 3/4 to 3 in. thicknesses, from 16 to 48 in. in width, and from 4'-0" to 10'-0" and longer in length. It can be used for roof, cavity wall, and perimeter insulation; as a base for plaster-ing; and as finish sheet or board materials (see Figures 30-1C and 30-2).

2. *Spray-on insulation.* This type is applied to surfaces by guns that mix and atomize the foam as it is being applied. It can be sprayed on the inside of buildings and on roofs, and penetrates inaccessible areas.

3. *Site-applied insulation.* This is hand-mixed and with special applicators. It can be used on site to insulate inaccessible areas such as spaces between walls and window and door frames, cavity walls, open areas in curtain wall construction, etc.

4. *Insulated sandwich panels* including doors, curtain wall units, etc. The foaming mixture is injected between skins of aluminum, steel, plastic, wood, plywood, etc., and because the foaming mixtures used are of the type that adheres and bonds to these skins, an extremely strong and rigid material is produced.

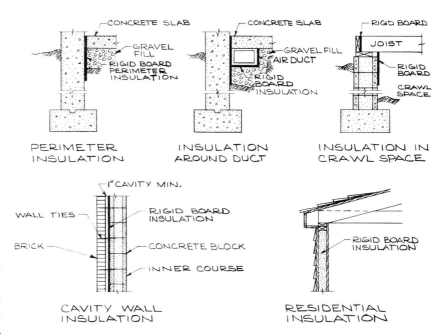

Figure 30-2. Applications of Rigid Board Insulation

Reflective insulation consists of sheets of highly reflective metal foil with air spaces between. It differs from other types of insulation because the number of reflecting surfaces, not the thickness of the material, determines its insulating value. To be effective, the foil must be exposed to an air space, because the only heat that can penetrate is radiant heat which can be reflected.

When so used, each reflecting surface is equivalent to approximately 1/2 in. of insulating board. When a reflecting metal foil is applied to a rigid material, it must be installed facing an air space, thereby adding to the total insulation (see Figure 30-1D).

Vapor Barriers

The formation of frost on the inside faces of windows in cold weather is an example of water vapor in a room condensing on the glass because of a temperature change. This can generally be eliminated by using storm sash, which actually is another means of insulation because of the introduced air space. Water vapor can pass through most building materials, and in cold weather it may condense within the wall structure. Such condensation may, in time, lessen the value of the insulation and also may cause paint and building materials to deteriorate. A vapor barrier, a membrane through which water vapor cannot pass, is essential to prevent moisture from reaching the insulation and causing condensation.

Metal foils make excellent vapor barriers, as do asphalt-treated felts, treated laminated papers, and polyethylene film of 0.002 or 0.004-in. thickness. The vapor barrier should be installed on the room side of the insulation, or behind the lath and plaster or other finish material.

Acoustical Materials

Sound. The science of controlling sound within buildings is called acoustics. Sound is a wave traveling in air. A dog barking across the street creates sound waves that travel through the air and to the ear. Sound is also transmitted by vibration through materials. Knocking on a pipe on one floor of a building will set up vibrations in the pipe that will travel up and down the pipe, creating sound waves in all the rooms where the pipe is located.

Types of acoustical materials. Selecting the acoustical material for effective sound absorption and transmission is becoming increasingly difficult, not only because of the many new building materials specif-

Table 30-2. *Classification of Acoustical Materials*

Types	Description
I	Regular perforated cellulose fiber tile
II	Random perforated cellulose fiber tile
III	Slotted cellulose fiber tile
IV	Textured, finely perforated, fissured, or simulated fissured cellulose tile
V	Membrane-faced cellulose fiber tile
VI	Cellulose fiber lay-in panels
VII	Perforated mineral fiber tile
VIII	Fissured mineral fiber tile
IX	Textured, finely perforated or smooth mineral fiber tile
X	Membrane-faced mineral fiber tile
XI	Mineral fiber lay-in panels
XII	Perforated metal pans with mineral fiber pads
XIII	Perforated metal lay-in panels with mineral fiber pads
XIV	Mineral fiber tile rated as part of fire-resistant assemblies
XV	Perforated asbestos board panels with mineral fiber pads
XVI	Sound-absorbent duct lining

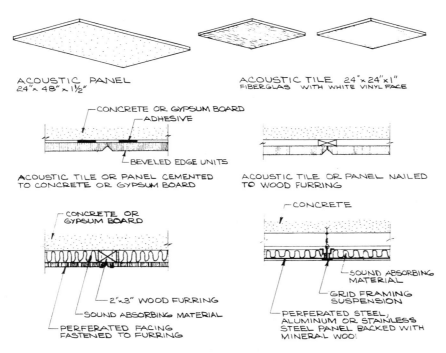

Figure 30-3. *Acoustical Materials and Types of Mountings*

ically designed to meet the demands for effective sound absorption and transmission but also because of the increasing knowledge in the science of acoustics. All building materials absorb sound and transmit it to some degree. Hard interior surfaces such as plaster and glass generally absorb less than 5% of the sound waves and reflect the balance. Effective absorption of sound is normally a characteristic of low-density materials, whereas dense materials are effective against sound transmission.

Sound-absorbing materials are relatively fragile and must be located so as to avoid abrasion and impact. This is one of the reasons for designing ceilings so that they assume a major role in the acoustical treatment of rooms. There are, basically, four broad categories of acoustical materials for sound absorption; namely (1) prefabricated, (2) "wet" or plastic-applied materials, (3) suspended baffles or "space" absorbers, and (4) special site-assembled materials.

Prefabricated or *factory-finished materials* include most of the products listed in the Acoustical Materials Association (AMA) bulletin (see Table 30-2). Basically these are factory-finished products, ready for installation.

Figure 30-3 illustrates some acoustical materials and the methods by which they can be mounted onto a ceiling. The acoustic tile or panel can be applied to a ceiling by means of adhesives, or nailed to wood furring strips.

Suspended acoustical ceilings are installed by providing various systems of grids, consisting of supporting channels or angles to which are attached various types of Ts, Zs, splines, etc., for supporting the tiles or panels (see Figure 30-4).

Wet or *plastic-applied materials* include acoustical plaster or mineral-type fiber products to which a binding agent is added at the time of application. They are applied in a wet or semiplastic state, either by

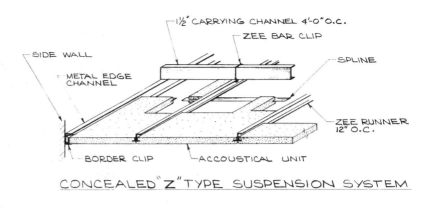

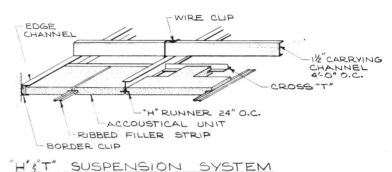

Figure 30-4. Suspension Systems for Acoustical Units

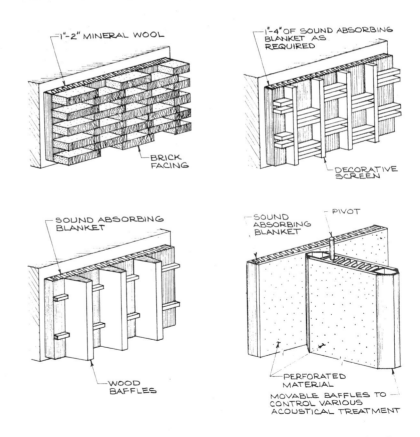

Figure 30-5. Architectural Materials Used for Sound Absorbing Treatments

253

hand troweling or machine spraying. Acoustical plasters are combinations of special aggregates, gypsum plaster, and additives to create air bubbles. As a rule acoustic plaster materials are covered by patents and trademarks.

Suspended baffles or *space absorbers* are prefabricated or specially designed sound-absorbing units such as flat baffles, cones, prisms, or crystalline shapes constructed of porous, sound-absorbing materials.

Special site-assembled acoustical materials and *treatments* are required when the interior architectural design is of a special kind — such as in a theater — or when the space requires special acoustical treatment — such as in a recording studio. A common method to achieve the required acoustical results is the use of a porous sound-absorbing material such as a 1 to 4-in. thick, sound-absorbing blanket placed behind a decorative open screen, or a series of wood baffles or grilles such as are shown in Figure 30-5.

REVIEW EXAMINATION

1. Materials that offer high resistance to the flow of heat are called what?

2. The more severe the heat or cold, the greater the need for what?

3. Flexible insulation is usually in what forms?

4. A roll or blanket-type of insulation material of 2-in. thickness has what U-factor (heat loss coefficient)?

5. What is a material that provides an effective moisture barrier?

6. The insulation with a U-factor of 0.16 is what?

7. Effective absorption of sound is normally accomplished by what type material?

8. There are basically how many broad categories of acoustical materials?

9. What is one method of applying acoustic tile to a concrete or gypsum board ceiling?

10. The effective control of sound transmission is accomplished by what type materials?

ASSIGNMENT

1. Make sketches showing three methods of fastening acoustic tile to a ceiling.

2. Illustrate by sketch two types of special site-assembled acoustical treatments.

SUPPLEMENTARY INFORMATION

Insulation of Pipes and Ducts

Pipes are insulated to prevent heat loss, condensation (dripping), and freezing, whereas ducts are insulated primarily to prevent heat loss or heat gain (see Figure 30-6). When water pipes are exposed to extreme cold, either in unheated portions of the structure or when built into exterior walls, they must be insulated to avoid freezing.

The dripping of water from pipes caused by condensation can be eliminated by insulation. Cold water pipes and pipes for the chilled liquid for air conditioning causes condensation and dripping and should be insulated in a similar manner. Heating pipes are insulated to prevent heat loss. The covering is made from cellular material in the form of a tube, split in half to allow for its installation. Rope-like insulations designed to be wrapped around the pipe are equally satisfactory. Coverings for ducts are in sheet or board form, solid or cellular.

Acoustical Treatments

The acoustical treatments for insulating against sound transmission through vibration are based on the use of dense materials such as lead and concrete or by means of isolation and vibration control.

Motors, machines, equipment, etc., which vibrate can be mounted on pads of thick lead, or layers of compressed cork, or be completely suspended by special isolation-control materials in order to control sound transmission.

Partition walls may be sound-insulated in a number of ways. Sound-absorbing material can be placed between staggered 2 × 4-in. wood studs, or double studs can be utilized, creating a separation (isolation) between the two stud partitions (see Figure 30-7). Two other methods are: (1) using a partition consisting of 2 × 4s and 2 × 2s with dense insulating board in the middle, and (2) using dense insulating boards as back-up for gypsum board.

Insulating Data

Types of commercial insulating materials

I. Materials used for insulation
  A. Fibrous
    1. Mineral wool
      a. rock wool
      b. glass wool
      c. slag wool
      d. mica
    2. Vegetable fiber
  B. Granular
    1. Mineral (vermiculite)
    2. Vegetable (granulated cork)
  C. Foam
    1. Urethane
    2. Polystyrene
  D. Reflective coatings applied to paper or other surfaces

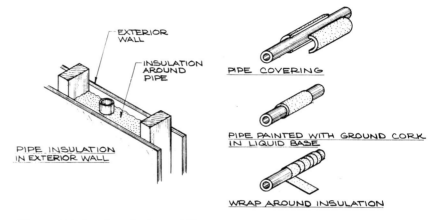

Figure 30-6. Insulation of Pipes

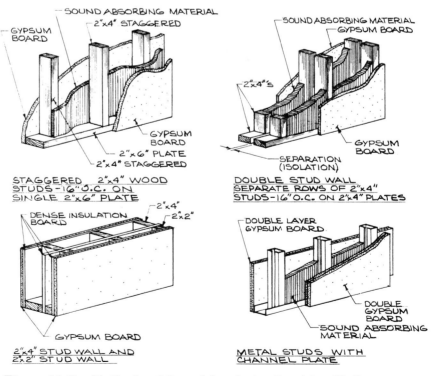

Figure 30-7. Methods of Sound Insulating Partition Walls

31

Roofing and Siding

INTRODUCTION

Roofing and siding are exposed to the damaging rays of the sun, rain, snow, sleet, ice, and wind. In order to make them weatherproof and waterproof, all of the natural elements have to be taken into consideration when roofing or siding material is selected and installed. This unit will deal with the various types of roofing and siding material, including wood, asbestos-cement, metal, asphalt, glass, built-up roofing, and plastic, as well as with the methods of installing these materials on buildings.

TECHNICAL INFORMATION

The climate where the building will be constructed must be carefully considered in selecting the type of roofing and siding materials to be used. For example, Miami has no snow, sleet, or ice and a short rainy season, but it also has long periods of hot clear days with very intense sunlight and a hurricane season with winds up to over 90 mph. Las Vegas has no snow, sleet, ice, or rain and winds up to only 40 mph., but it has almost continuous days of intense sunlight.

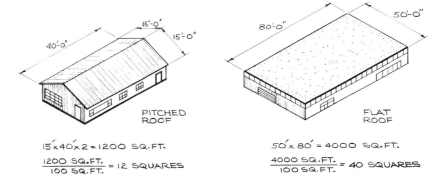

Figure 31-1. Roofing Material Units

Portland, Maine has rain, sleet, large quantities of snow and ice, winds as high as 60 mph., and only a month of intense bright sunlight.

Roofing

Roofing may be defined as the method or materials used to make a

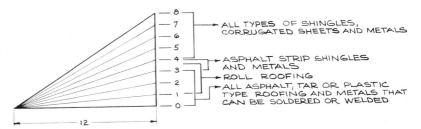

Figure 31-2. Diagram Showing the General Types of Roofing Materials Used for Various Roof Pitches

257

Table 31-1. *Types of Materials Used for Roofs With Pitches Over 3 to 12*

Type of Material	Type of Roofing
Aluminum	Corrugated, shingles, and sheet
Asbestos-cement	Corrugated and shingles
Asphalt	Shingles and strip shingles
Clay tile	Shingles
Galvanized steel	Corrugated and sheet
Copper, lead, stainless steel, and ternplate	Sheet
Corrugated wire glass	Corrugated
Roll roofing	Rolls
Slate	Shingles
Wood	Shingles

Table 31-2. *Types of Materials Used for Roofs With Pitches Less Than 3 to 12 to Flat*

Type of Material	Type of Roofing
Asphalt or tar	Built-up, 10- to 20-year guarantees
Canvas	Rolls
Copper, lead, stainless steel, and ternplate	Rolls
Plastic	Rolls, liquid, foam, and paint
Roll roofing	Rolls

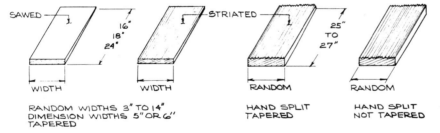

Figure 31-3. Types of Wood Shingles

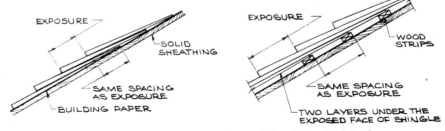

Figure 31-4. The Two Methods of Installing Wood Shingles

Table 31-3. *Exposure of Wood Shingles for Various Roof Pitches*

Type of Shingle	Length	Exposure for 5 to 12 Pitch and Greater	Exposure for 3 to 12 to 4 to 12 Pitches
Dimension and random	16"	5"	3-3/4"
	18"	5-1/2"	4-1/4"
	24"	7-1/2"	5-3/4"
Hand split	25" to 27"	7-1/2" to 8"	Cannot be installed

roof weathertight and waterproof. All types of roofing materials are figured by the *square*, which is 100 sq. ft. (see Figure 31-1).

Roof pitch. Roofs upon which roofing materials are applied can be divided into two main categories: (1) Roofs with pitches of 3 to 12 and greater; and (2) Roofs with pitches less than 3 to 12 ranging down to flat (see Figure 31-2). The materials used for the two categories are shown in Tables 31-1 and 31-2.

Roofs with pitches of 3 to 12 and greater quickly drain off the water from rain, snow, and sleet; therefore they do not have to be completely waterproof. In like manner, the steeper the pitch the less chance there is for high winds to lift off the materials except under extreme conditions, such as during hurricanes. In climates where there will be heavy snowfall it is advisable to use pitches of 6 to 12 and greater so that the snow will immediately slide off and not collect on the roof in large quantities and then, in a thaw slide off and create a hazard for people below. For roofs in this category, only the intense sunlight and the wind must be carefully considered.

Shingles

Shingles are manufactured as single units or in units simulating single shingles from 3'-0" up to 6'-0" in length. These various types are manufactured of wood, slate, clay tile, asbestos-cement, asphalt, and aluminum.

Wood shingles are single shingles and are manufactured in three types: dimension, random, and hand split (see Figure 31-3). They are installed either on solid sheathing or on wood strips (see Figure 31-4). The exposure varies with the roof pitch (see Table 31-3).

Slate may be laid on wood strips spaced the same as slate exposure or on solid sheathing with 15-lb. felt first applied. Slate colors are blue-gray, black, gray, purple, green, and red (see Figure 31-5).

Clay tile roofing (shingle) can only be used on roofs with pitches 4-1/2 to 12 and greater. Clay tile shingles are made from the same clay as used for bricks and manufactured by the soft-mud and dry-press methods. They are generally a red color but can be ordered with colored glazes in a wide variety of colors (see Figure 31-6). Clay tile roofing is manufactured in various types, such as Spanish, Mission, French, and English.

Asbestos-cement shingles are made from portland cement and asbestos fibers, which act as reinforcing. They are available either plain or striated or with a wood textured surface, and in a variety of colors. Roof pitches 3 to 12 and greater are required (see Figure 31-7). There are two other types of asbestos-cement shingles — Dutch lap and hexagonal, which give a different design to the roofing.

Slate, clay tile, and asbestos-cement shingles are very heavy materials and therefore the supporting structure should be designed to receive this extra load. Asbestos-cement shingles vary in weight from 265 to 585 lbs. per 100 sq. ft., slate weighs 700 to 810 lbs. per 100 sq. ft., and clay tile weighs 900 lbs. per 100 sq. ft.

Asphalt individual and *strip shingles* are manufactured in various colors and in weights ranging from 210 lbs. to 345 lbs. per square (100 sq. ft.), and also in various degrees of fire resistance. They are installed on roofs with pitches of 3 to 12 and greater, and they must be installed on solid sheathing. In order to overcome high winds they can be obtained with special tabs of adhesive near the butt end (see Figure 31-8). Single asphalt shingles are manufactured in hexagonal types and also in a Dutch lap type.

Roll Roofing

Asphalt roll roofing is manufactured in rolls 36 in. wide and in

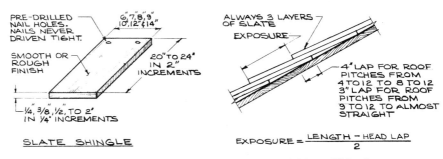

Figure 31-5. *Size, Shape, and Installation of Slate Shingles*

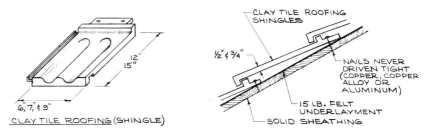

Figure 31-6. *Size, Shape, and Installation of Clay Tile Roofing (Shingles)*

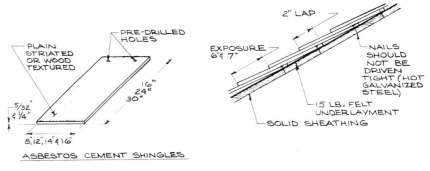

Figure 31-7. *Size, Shape, and Installation of Asbestos Cement Shingles*

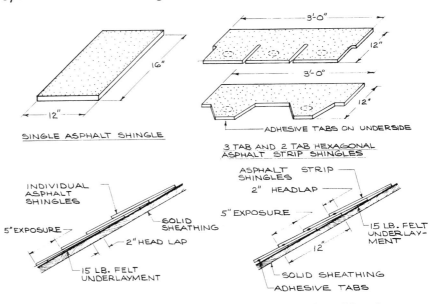

Figure 31-8. *Size, Shape, and Installation of Asphalt Shingles*

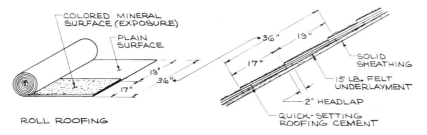

Figure 31-9. *Size, Shape, and Installation of Roll Roofing*

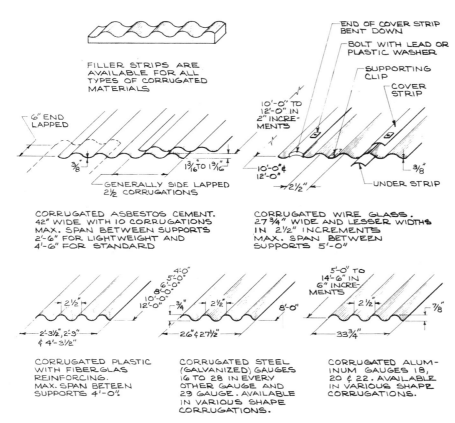

Figure 31-10. *Corrugated Roofing Materials*

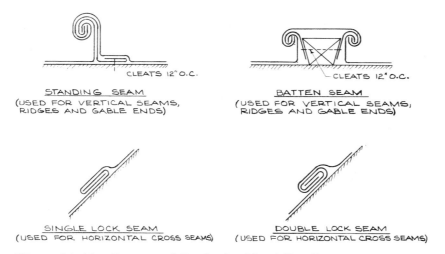

Figure 31-11. *Seams and Locks for Metal Roofing*

lengths of 36'-0" to 48'-0". The exposure surface is a colored mineral surface and the remaining surface is plain. This roofing is used for roof pitches from 1 to 12 to 4 to 12 (see Figure 31-9).

Corrugated Roofing

Corrugated asbestos-cement, wire glass, plastic, and metal are used for roofs with pitches of 3 to 12 and greater. They are joined at the sides by lapping the corrugations (except wire glass) and are generally end-lapped 6 in. They all require pre-drilling for holes (except the plastic type), and all methods of attaching should allow slight movement; therefore, they are never tightly attached (see Figure 31-10).

Both corrugated steel and aluminum are manufactured in various different shapes of corrugation. There are two types: *standard* and *industrial*, in which only the gauges and the depth of corrugations change.

Metal Roofing

The metals, copper, lead, stainless steel, aluminum, and ternplate are used for roofs with pitches of 3 to 12 and greater. The methods of installation are similar, and attachments, screws, and nails must be of a type compatible with the metal used for roofing. For example, copper nails are to be used with copper roofing and aluminum nails with aluminum roofing (see Figure 31-11). Roofs with pitches ranging from 3 to 12 to flat require that wind, waterproofing, drainage, snow, sleet, ice, and intense sunlight must be carefully considered.

The metals that can be easily soldered or welded are used for roofs that are flat and up to 2 to 12 pitches. The metals used are copper, lead, zinc, ternplate, and stainless steel (see Figure 31-12).

When using any metal for roofing, care should be taken that all the materials in contact with the metal or used for installing the metal will not set up a galvanic action that will corrode or disintegrate the metal roofing or the materials in contact with metal or the nails, screws, clips, etc. used in installing the metal — for example, aluminum roofing with stainless steel gutter (see Figure 31-13).

Built-up Roofing

Asphalt and *tar built-up roofing* consists of alternate layers of either asphalt- or tar-saturated felts (paper) and hot or cold asphalt or hot tar (known as pitch). Each layer of felt and asphalt or tar is called a ply. This type of roofing is designated as 3-ply, 5-ply, etc. The first layer is not considered a ply because it consists of one or two layers of unsaturated felt, nailed down to the substrate. This layer is of utmost importance because it is the method of bonding (holding down) the built-up roofing to the substrate. Generally a 4-ply built-up roof consists of 1 ply of unsaturated felt and 4 plies of saturated felt and pitch. To protect the plies from intense sunlight, a finish surface of slag or stone chips is applied. Built-up roofing is applied to roofs that are flat and with pitches up to 2 to 12 (see Figure 31-14).

Canvas has been used successfully for residential roof terraces and decks for many years. The solid sheathing is given a heavy coat of white lead and the canvas is applied in strips lapped 2 in. at joints and secured with copper tacks at lapped edges. When installed, two coats of exterior oil base paint are applied.

Plastic types of roofing were first developed to meet the new roof forms designed with reinforced concrete, such as thin shell domes. These new roof shapes were not adaptable to the then-existing types of roofing materials.

Plastic roofing materials are manufactured as single-ply, singleply combined with foam-type insulation, spray-on, roll-on, trowel-on

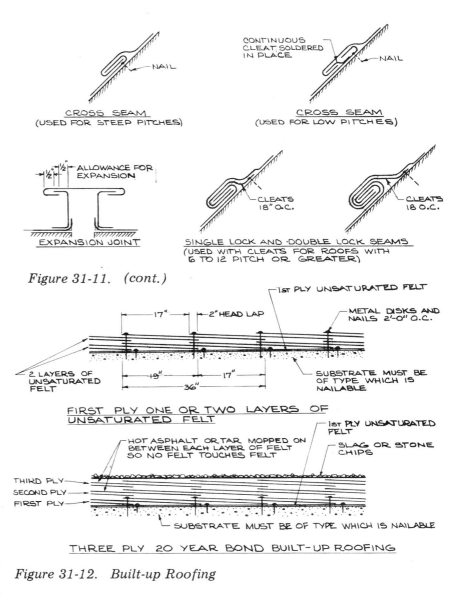

Figure 31-11. (cont.)

Figure 31-12. Built-up Roofing

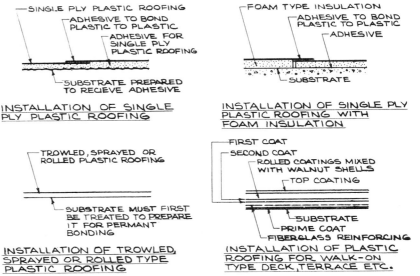

Figure 31-13. Installation of Various Types of Plastic Roofing

261

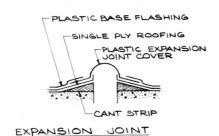

EXPANSION JOINT

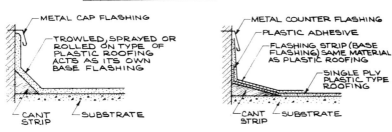

SPRAY-ON, TROWLED OR ROLLED PLASTIC ROOFING AT VERTICAL WALL

SINGLE PLY PLASTIC ROOFING AT VERTICAL WALL

Figure 31-14. Flashing and Expansion Joints Used with Plastic Types Roofing

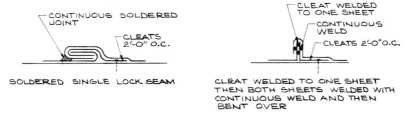

SOLDERED SINGLE LOCK SEAM

CLEAT WELDED TO ONE SHEET THEN BOTH SHEETS WELDED WITH CONTINUOUS WELD AND THEN BENT OVER

Figure 31-15. Methods of Joining Metal Roofing on Flat or Roofs with Pitches up to 2 to 12

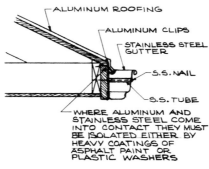

Figure 31-16. Detail of Stainless Steel Gutter and Aluminum Roofing

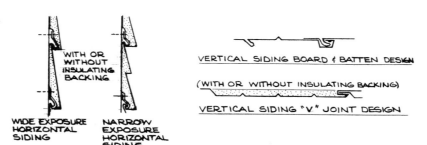

WIDE EXPOSURE HORIZONTAL SIDING

NARROW EXPOSURE HORIZONTAL SIDING

VERTICAL SIDING BOARD & BATTEN DESIGN

(WITH OR WITHOUT INSULATING BACKING)

VERTICAL SIDING "V" JOINT DESIGN

Figure 31-17. Aluminum Siding

and brush- or mop-on. In all cases the type of substrate must be checked so that the type of plastic roofing to be installed will permanently bond (adhere) to the substrate. In general, a substrate that has a high moisture content or high absorption rate requires special treatment when a plastic type roofing is to be applied (see Figure 31-15).

These roofing materials have led to the development of a completely new set of flashing and expansion joint details (see Figure 31-16). In general, most plastic-type roofing materials are neoprene, silicone, synthetic rubbers, urethane, and adhesives based on the same material as the roofing. They are manufactured in various sunfast colors, and all the various flashing and expansion joint materials and devices are available.

Siding

Many roofing materials are also used for siding, among them wood, slate, asbestos-cement, and asphalt shingles. Types of siding materials include those which are applied over a sheathing, prefabricated panels including exterior facing, insulation, and interior finish (see Table 31-4).

Aluminum siding. Aluminum siding is manufactured in various types: horizontal (wide and narrow design) and vertical (V-joint or board-and-batten design). It is available 0.025 in. thick, in a wide variety of colors, either baked on or a synthetic type paint finish (see Figure 31-17).

Aluminum industrial siding is manufactured in sheets from 3'-0" to 30'-0" in lengths, in widths of 3'-5-5/8", 1'-7-1/2", and others depending on manufacturer. These types of siding are available as prefabricated panels with insulation and interior finish (see Figure 31-18).

Asbestos-cement siding. Asbestos-cement panels are manufactured

262

in sheets 1/8 and 3/16 in. thick, 4'-0" wide, and 4'-0", 8'-0", 10'-0", and 12'-0" long, in various colors. They must be predrilled for nailing (see Figure 31-19).

Asbestos-cement siding in large single strips 12" × 4'-0" long are predrilled and installed the same as shingles. They are manufactured in a wide range of colors.

Asbestos-cement sandwich panels consisting of an insulation-type board with two sheets of 1/8-in. asbestos-cement panels bonded to both sides are used for checking, curtain walls, and exterior walls for residences.

Steel siding. This is manufactured in many shapes similar to aluminum, with various types of protective surface finishes. These panels can be single sheet or a combination of sheet and backing insulation, or a complete prefabricated panel with exterior and interior finish having an insulating material sandwiched between.

Wood siding. Wood is manufactured in horizontal siding in various shapes and in boards for vertical siding (see Figure 31-20).

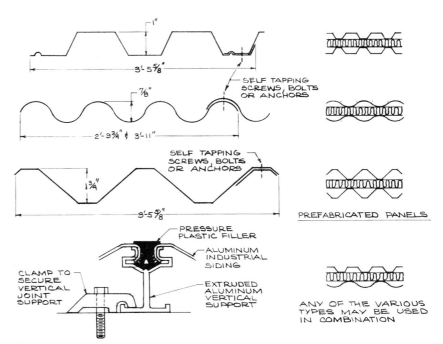

Figure 31-18. Industrial Aluminum Siding Shapes, Sizes, and Methods of Attaching onto Materials of Buildings

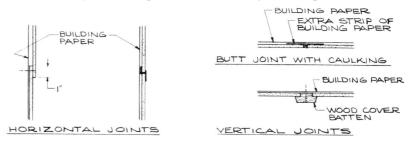

Figure 31-19. Installing Asbestos Cement Sheets as Siding Material

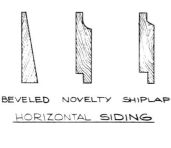

BEVELED   NOVELTY   SHIPLAP

HORIZONTAL SIDING

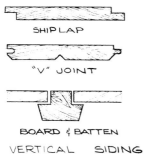

SHIPLAP

"V" JOINT

BOARD & BATTEN

VERTICAL SIDING

Figure 31-20. Types of Wood Siding

Table 31-4. Siding Materials

Materials	Shingles as Siding	Siding
Aluminum	Shingles	Siding, and industrial siding and prefabricated panels
Asbestos-cement	Shingles	Siding, panels, and prefabricated panels
Asphalt	Shingles	Various textured siding materials
Plastic		Fiberglass panels
Slate	Shingles	
Steel	Shingles	Siding with special colored protective coatings and prefabricated panels
Wood	Shingles	Siding and boards

263

1. Roofs with pitches from 3 to 12 and greater are required to be what?

2. Roofs with pitches from 3 to 12 to flat are required to be what?

3. The climate of the area where a building is to be constructed should be checked for what six things before selecting roofing material?

4. What are the two categories of roofs upon which roofing materials are applied?

5. What are the three types of wood shingles?

6. When installing slate, clay tile, or asbestos-cement shingles on a roof, what should first be considered regarding the supporting structure?

7. What should asphalt strip shingles have to overcome high winds?

8. Asphalt roll roofing can be installed on roofs with what degree of pitch?

9. Corrugated type roofing materials other than glass are joined at their sides by what?

10. Metals, when used for roofing on roofs with pitches 3 to 12 and greater are joined at vertical seams, ridges and gable ends with what?

11. Metals, when used for roofing on roofs with pitches of 3 to 12 and greater are joined at horizontal or cross seams with what?

12. Built-up roofing consists of layers called what?

13. Plastic-type roofing materials are installed by what three methods?

14. When using metals for roofing materials, why should care be taken with materials in contact with the metal or used for installing the metal?

15. What qualifications must metals used for roofs that are flat and up to 2-to-12 pitches have?

16. Name the roofing shingles that are also used as siding.

17. Is metal siding, either aluminum or steel, manufactured as complete prefabricated panels with inside and outside finishes and insulation?

ASSIGNMENT

1. Draw a freehand diagram of roof pitches ranging from 0 to 12 to 8 to 12 and list the various types of roofing material used for various roof pitches.

2. Draw freehand, to scale, the flashing and expansion joint details for plastic-type roofing materials.

SUPPLEMENTARY INFORMATION

Substrate. This is the term to describe the material upon which roofing materials are to be applied.

Wood shingle designations. There has been established by shingle manufacturers a system to designate the thickness of wood shingles (see Figure 31-21).

Slag. A nonmetallic waste product obtained from the smelting of ferrous metals, slag is the remaining material from the metal ores after the metal has been extracted. In granular form it is used as topping for built-up roofing, and in powdered form it is used in masonry cements.

Copper roofing. When copper is used as a roofing material, care should be taken that any water running or dripping from it will not come in direct contact with other materials, because copper, when exposed to the weather, will develop a bright green protective coating that can run and stain other materials.

TCS roofing. Terne-coated stainless steel is a nickel-chrome stainless steel covered on both sides with terne alloy (80% lead, 20% tin). It does not need maintenance if properly installed and has proven itself for decades; it will outlast virtually any building on which it is applied.

The terne coating on TCS is anodic to stainless steel, which means that the terne will sacrifice itself to protect the core metal. It is therefore recommended for use in severe industrial and chemical environments.

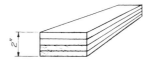
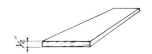

THE BUTT OF FOUR (4) SHINGLES = 2"
DESIGNATION 4/2

THE BUTT OF ONE (1) SHINGLE = ½"
DESIGNATION 1/½

Figure 31-21. Thickness Designations of Wood Shingles

32

Doors and Windows

INTRODUCTION

This unit deals with the various types of wood and metal windows such as double-hung, projected, casement, awning, and others. Details of installation, architectural symbols, glazing, and the various finishes available for wood or metal windows are discussed. This unit also deals with the various types and kinds of doors, their installation, hardware, architectural symbols, and operation identifications.

TECHNICAL INFORMATION

Windows

Windows originally were used for admitting daylight and later for ventilation. Perhaps the first ventilating window was the casement type because it required the simplest type of hardware — only pivots, a simple latch, and a bar of metal to hold it open. Today we have a great variety of window types and with the increasing use of air conditioning, the provision for ventilation is becoming almost unnecessary, and the window is returning to what it originally was — a means of admitting daylight.

Generally all wood or metal stock windows are standardized as to sizes and are manufactured as complete units that include hardware, weatherstripping, operating mechanism, etc., come glazed or unglazed, and are ready to be installed in a building. They are manufactured to receive window glass, 1/4-in.-thick plate, 1/4 in. special glass, or insulating glass. They also may be obtained with screens and storm sash.

Windows are manufactured in four categories based on their use in building construction: residential, commercial, industrial, and monumental. All types have established standards that have been developed by window manufacturers' associations and institutes, U.S. Department of Commerce, American Standards Association, and other city, state, and federal agencies.

In architectural elevation drawings (see Figure 32-1), standard symbols are used to show the various types of ventilation (window swings).

Test data are available for all types of wood or metal windows — included are horizontal and vertical

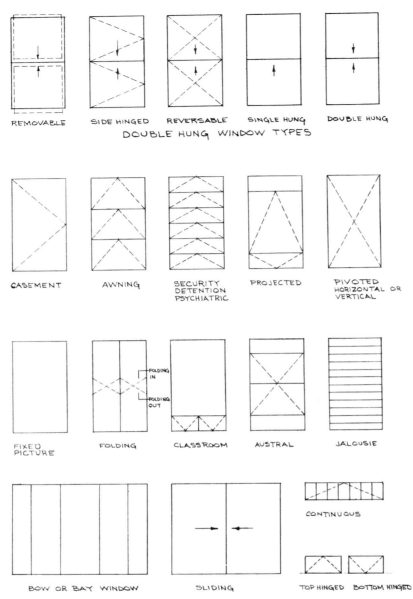

Figure 32-1. Window Types and Standard Architectural Symbols to Show Types of Ventilation Operation

Table 32-1. Factory Finishes and Permanent Finishes for Windows

Type of Window	Factory Finish	Permanent Finish Job- or Factory-Applied
Wood	Preservative against fungi, insects, and water	Paint, stain, vinyl, or aluminum
Steel	Shop prime coat or corrosion-resistant prime coat	Paint, baked enamel, vinyl
Aluminum	Protective coating	Smooth to high polish, colored anodized finishes, and baked enamel
Stainless steel	Protective coating	Colored, smooth to high-polish finishes

loads, air infiltration and wind, torsion tests for projected and awning types, and water resistance. When a window type and grade have been selected to meet the design requirements in a building, obtaining test data and details from various manufacturers of this type and grade of window enable the selection of the correct type, grade, and manufacturer.

All types of windows, when assembled and ready to be delivered to a building under construction or to warehouses and distributers, receive a factory-finishing treatment. This treatment may be either protective, preservative, permanent finish, or a base for painting, staining, or other type of finish (see Table 32-1).

Wood windows. Standards governing the production of wood windows have been established for double-hung, casement, and awning windows. These standards cover materials, sizes, frame, sash, check rails, hardware, operating mechanisms, weatherstripping, and screens and storm sash. The standards require that all wood parts receive preservative treatment against fungi, insects, and water.

Other types of wood windows include sliding, picture (fixed), bay or bow window, modular units, cellar windows, and various other types and operating systems.

Most wood windows are manufactured as complete units ready to be installed, but are also available as separate sash and frames. They may be obtained with screens, storm sash, single glaze, double glaze or with insulating glass.

Ponderosa pine, kiln-dried to a moisture content of 6 to 12% is the wood generally used in the manufacture of wood windows. Of all wood window types, the double-hung window is the most widely used. The details in Figure 32-2 show an elevation of a double-hung wood window indicating that details were taken at the head section

(A), the meeting rail (B), the sill section (C), the muntin (D), and the jamb section (E). Although the sash for most residential windows is 1-3/8 in. thick, commercial and institutional sash are 1-3/4 in. thick. Insulating glass 1/2-in. thick can be used with 1-3/4-in. sash; however, 2-1/4-in. sash is recommended for 3/4 and 1-in. insulating glass because these thicknesses are too heavy for 1-3/4-in. sash.

Glazing for wood windows should be done with a wood glazing-type compound, knife consistency, or extruded plastic-type gaskets. All glass should be bedded with the same material if a glazing compound is used (see Figure 32-3). All exterior nails that cannot be set should be aluminum or stainless steel, and all other exterior nails that can be set and puttied should be hot-dipped, galvanized steel or aluminum.

Steel windows. All types of steel windows are manufactured as a complete unit including hardware, weatherstripping, and operating mechanism, but they are always site-glazed.

Standards have been established regarding sizes, material, weatherstripping, hardware, glazing, and operating mechanism for residential, commercial, industrial, and monumental windows. These standards are based upon use. For example, the windows used for an industrial building do not have to meet the rigid requirements that are established for a hospital or a library.

Figure 32-4 shows the installation of a steel window in a typical commercial building. Note the anchor fasteners used at head, jamb, and sill.

Because all steel windows are to be site-glazed, the correct method of glazing should be selected. Steel windows are manufactured with interior or exterior glazing with or without stop beads (see Figure 32-5).

Aluminum windows. Aluminum extrusions are used to manufacture all the various types and grades of aluminum windows. The alloys most generally used are 6063-T5

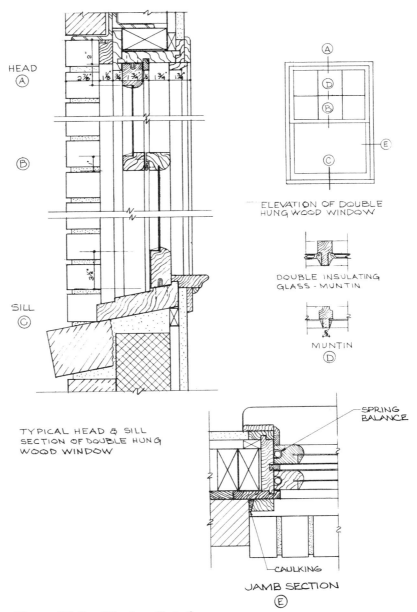

Figure 32-2. Window Details

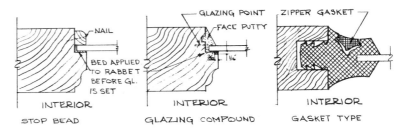

Figure 32-3. Glazing Details in 1-3/8" Sash

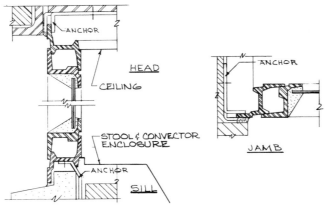

Figure 32-4. *Installation of Steel Window in a Typical Commercial Building*

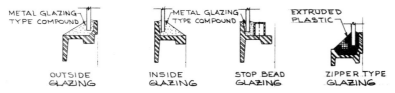

Figure 32-5. *Glazing for Steel Windows*

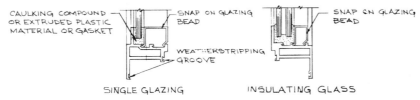

Figure 32-6. *Snap-in Glazing Beads*

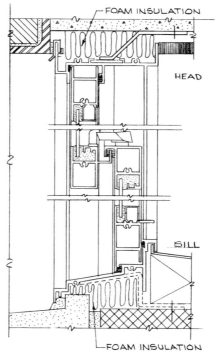

Figure 32-7. *Detail of an Aluminum Double Hung Window*

and 6063-T6. These alloys have tensile strength in excess of 22,000 PSI, which is the minimum set by the Architectural Aluminum Manufacturer's Association (AAMA). A minimum of 0.062 in. has been established for material or section thickness. Joints are either welded or fastened by mechanical means. All hardware, operating mechanism, weatherstripping, fasteners, anchors, etc., must be of a metal that is compatible or isolated from aluminum so that no galvanic action can occur. Aluminum-to-aluminum contact, where parts move against one another, is not permitted.

Extrusions allow for various small grooves, projections, etc. to be formed, which make possible a snap-on type of glazing stop-bead (see Figure 32-6).

Details of an aluminum double-hung window installed in a building of reinforced concrete with cavity wall construction are shown in Figure 32-7.

Stainless steel windows. A limited number of manufacturers produce types and grades of stainless steel windows. Stainless steel is available in various finishes ranging from highly polished to a dull sand-textured finish. It is highly corrosion-resistant and exceptionally strong in thin gauges. Stainless steel sheets in guage Nos. 16, 18, 20, and 22 of alloy types 202, 302, and 430 are generally used. Figure 32-8 shows a jamb section of an awning type window.

As with other metals, all hardware, operating devices, fasteners, anchors, etc. should be of a metal compatible with stainless steel so that no galvanic action can occur.

Doors

Doors are primarily used for exterior entrances and exits from a building and to close off or subdivide areas within a building. The materials most generally used in the manufacturing of doors are wood, steel, aluminum, stainless steel, glass, fabrics, and plastic. Doors can be identified by their method of opening, such as swinging, folding, sliding, bypassing, overhead, revolving, and rolling (see Figure 32-9).

The most widely used doors are the swing type. To establish a uniform method of denoting the swing and establishing a set of standards, the hardware manufacturers set specific names that are used when facing swinging doors (see Figure 32-10).

In general, doors are site-installed (although several types are available as a complete preassembled unit including frame, door, and hardware) and therefore it is necessary to select the type of finish hardware to be used such as butts (hinges), pulls, push plates, knobs, latches, locks, closers, etc. The various locations and dimensions for finish hardware have been standardized, as shown in Figure 32-11.

Building codes require certain doors to be fire-resistant. Doors are available in various ratings such as 1/2-hr., 1-hr., etc., to meet these fire-rating requirements.

Wood doors. Wood doors are manufactured for both interior and exterior use. They are available in the following standard dimensions: 1-3/8 and 1-3/4-in. thickness, 6'-8" and 7'-0" in height and widths, from 1'-10" to 3'-0" in 2-in increments. Exterior doors are generally 1-3/4 in. thick, made with waterproof adhesives, and treated against decay, fungi, and insects, whereas interior doors generally are 1-3/8 in. thick (1-3/4 in. where heavy use will be encountered), made with nonwaterproof adhesives, and not treated for decay, fungi, and insects. Wood doors are also manufactured with fire ratings, acoustical treatment, and X-ray shielding.

Figure 32-12 shows various wood swing-type doors. They are manufactured in different ways such as with solid wood, solid wood core

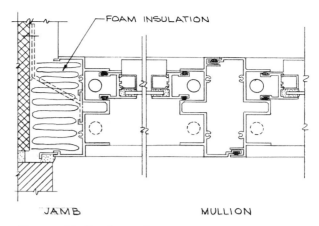

Figure 32-7. (cont.)

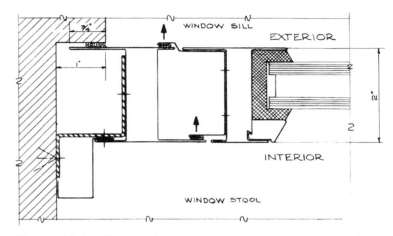

Figure 32-8. Jamb of Stainless Steel Awning Type Window

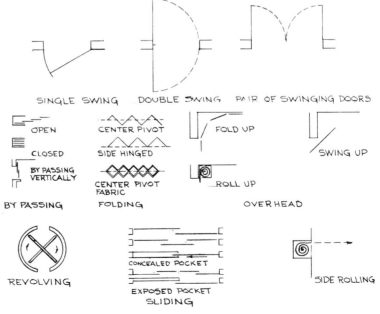

Figure 32-9. Types of Doors Identified by Their Opening Action

271

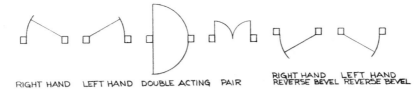

Figure 32-10. *Specific Names Used when Facing Swing Type Doors*

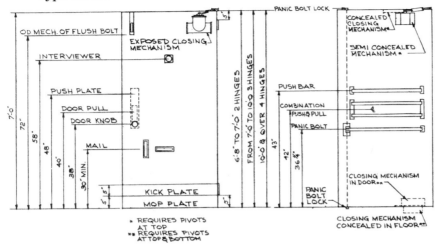

Figure 32-11. *Location of Typical Finish Hardware for All Types of Doors*

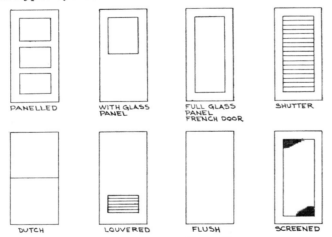

Figure 32-12. *Wood Swing Type Doors*

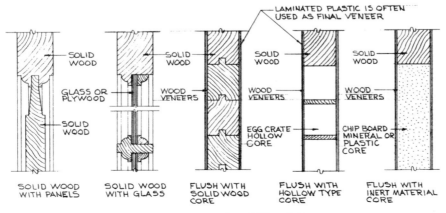

Figure 32-13. *Construction Details of Wood Doors*

with wood veneers, hollow core with wood veneers, and various core materials with wood veneers (see Figure 32-13).

Wood constructed doors are used for folding (flush or panelled), overhead fold-up (flush or panelled), sliding (flush or panelled), bypassing (flush or panelled), and side-rolling (small strips of solid wood) type of doors.

Steel doors. The greatest use of steel doors in building construction is for interior doors requiring fire ratings and for entrance and exit doors with or without fire ratings. Fire ratings are as follows: *class A* 3 hrs; *class B* 1-1/2 hrs., *class C* 3/4 hrs., *class D* 1-1/2 hrs., and *class E* 3/4 hr. Steel doors are available in the same types as swing-type wood doors (see Figure 32-12) and also as overhead roll-up, fold-up and swing-up, sliding, revolving, and bypassing. They are usually manufactured of 14, 16, 18, and 20-gauge cold-rolled steel with all joints welded and ground smooth; 10, 12, and 14 gauges are used for reinforcing channels and angles and as reinforcing for hinges, closures, locks, etc. (see Figure 32-14).

Aluminum doors. In building construction, aluminum doors are used for entrance doors, which include the frames, and for interior swing-type doors where no fire ratings are required. They are used for swing-type, bypassing, revolving, overhead, and sliding doors.

They are generally manufactured from 6063-T5 aluminum alloy; joints are welded or mechanically secured and meet standards established by the National Association of Architectural Metals Manufacturers; they are available in the same types as wood swing-type doors (see Figure 32-12).

Figure 32-15 shows details of a typical aluminum entrance door, transom, and side lights.

Aluminum doors for interior work are manufactured by various different methods, as shown in Figure 32-16.

Glass doors. Tempered glass is used for swing-type doors and revolving doors. This kind of glass requires that all pivots, locks, holes, etc., be part of the casting when the tempered glass is manufactured. Figure 32-17 shows a typical elevation and section of a tempered glass door.

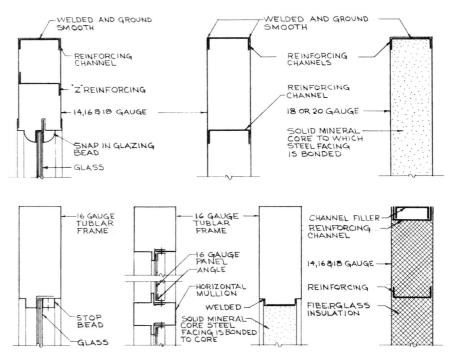

Figure 32-14. *Various Types of Steel Door Construction*

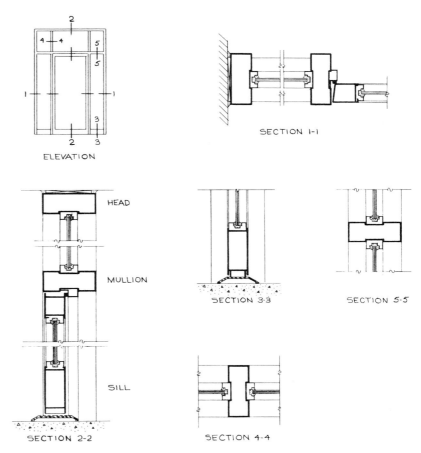

Figure 32-15. *Aluminum Entrance Door with Transom and Side Lights*

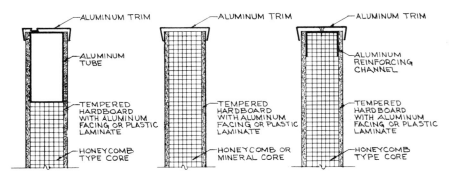

Figure 32-16. Interior Aluminum Doors

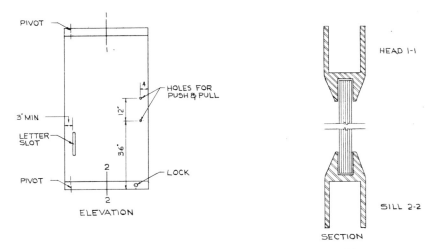

Figure 32-17. Typical Tempered Glass Swing Type Door

REVIEW EXAMINATION

1. Name the four window categories by their use.

2. Are the vast majority of wood windows (a) double-hung, (b) casement, or (c) projected?

3. What is the wood almost universally used for wood windows?

4. What is the moisture content permitted in standard wood windows?

5. Are steel windows delivered to a building construction site glazed or unglazed?

6. Do the aluminum window specifications of the Architectural Aluminum Manufacturer's Association require a minimum tensile strength of (a) 22,000 PSI, (b) 27,000 PSI, or (c) 35,000 PSI?

7. What is a minimum required material thickness for aluminum windows?

8. Name four types of doors identified by their opening action.

9. Do swing type doors have specific names used to denote the swing?

10. Are there standards for locating finish hardware for doors?

11. Wood doors are manufactured in what two standard thicknesses?

12. Are steel doors labeled for fire rating?

13. What is the aluminum alloy used in the manufacturing of doors?

1. Sketch five door types in elevation.

2. Sketch five window types in elevation showing swing designation.

3. Sketch the symbolic representation of the following doors:

(a) Swing, two types.

(b) Folding, three types.

(c) Overhead, three types.

SUPPLEMENTARY INFORMATION

Door Frame and Door Buck

All doors have a frame into which they are hung or installed. This is known as the door frame or door buck. The buck must not only be of the size and type to meet the architectural design requirements but also must meet the requirements for the joining of the door and frame to the materials into which it is to be installed.

As a general rule, the buck is of the same material as the door — a wood door requires a wood buck, a steel door a steel buck, etc. A significant change to this rule must take into consideration that door and buck must be of compatible materials so that no chemical or galvanic action can occur. For example, a steel buck with an aluminum door for exterior entrances would set up a galvanic action whereby the aluminum would be destroyed.

Figure 32-18 illustrates various types of wood bucks with wood doors. Two of the jamb sections are for interior doors where plaster board is shown on both sides and where ceramic tile on plaster board on one side with wood panelling on the other.

The exterior jamb sections include insulation between the walls, outside sheathing, and siding or brick veneer.

Figure 32-19 shows various types of steel bucks with wood or steel doors.

The jamb sections shown in Figure 32-19 indicate metal bucks

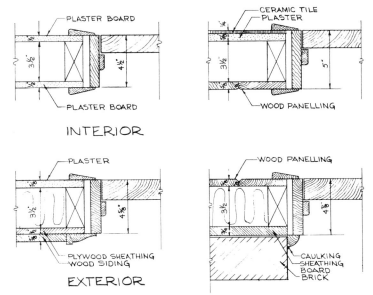

Figure 32-18. Typical Wood Door Frames (Bucks)

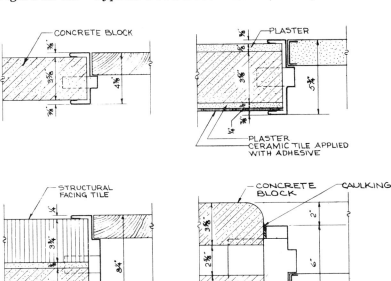

Figure 32-19. Jamb Details of Steel Door Bucks

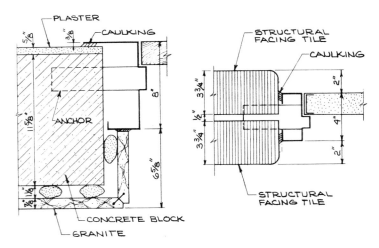

Figure 32-20. Jamb Details of Aluminum Door Bucks

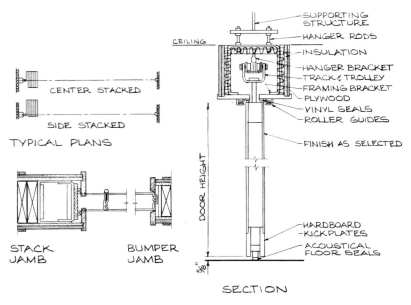

Figure 32-21. Folding Partition

anchored into a concrete block wall, a concrete block wall with plaster finish and ceramic tile applied with adhesives, concrete block with structural facing tile, and an exterior wall with brick and concrete block.

Figure 32-20 illustrates aluminum door frames or bucks and how they are set into a structural facing tile wall and into a solid masonry wall with granite facing.

Figure 32-21 shows a stack and bumper jamb and a head section through a folding partition. These partitions are available with manual operation or electrical motorized operation. While the partition is in motion a 5/8-in. clearance is allowed between the bottom of the partition and the finished floor. The acoustical floor seals are automatically actuated by the final closing action of the partition and will also lock the partition firmly in place without the use of floor bolts or floor locks.

REVIEW EXAMINATION — UNITS 25-32

1. What must the minimum clearance between the edges of plate glass and its frame be?

2. Before glass is installed into a wood frame, how should the wood be treated?

3. What is elastic glazing compound made from?

4. Glass block walls cannot be used for what type of walls?

5. When glass block is installed in a masonry or steel frame opening, what is the size limitation?

6. Heat-strengthened glass is generally used in what kind of walls?

7. What type plastic is used in the manufacture of skydomes?

8. Vinyl flooring is manufactured in rolls and in tile form. What sizes of square tiles are available?

9. Translucent corrugated plastic sheets are used in the construction field for what purpose?

10. Polyurethane and polystyrene plastics produce a pre-formed and foamed type of insulation. What is

the U-factor for 1-in.-thick preformed plastic insulation?

11. What is the maximum size manufactured for a silvering, mirror-glazing type of plate glass?

12. What is the liquid part of the paint known as?

13. What is the solvent for shellac?

14. What is the hiding power of the type of white paint solid known as zinc oxide?

15. The best results in painting are achieved when the paint is applied between what temperatures?

16. What does the word "length," when used in relation to varnish, mean?

17. What do the plate-like particle shapes in white paint solids (pigments) increase?

18. What is one method of cleaning ferrous metals?

19. Which type of metals receive a protective removable coating that is taken off when the possibility of damage or staining is no longer present?

20. What must be provided when 9 × 9 in. wood block flooring is installed directly onto a concrete slab on grade?

21. In order to provide for the expansion and contraction of terrazzo flooring, what is it necessary to do?

22. Where in a building should terrazzo flooring be installed?

23. What must the surface be upon which ceramic and quarry tile flooring is to be installed?

24. When concrete is to be used as the finish floor surface, what must it be treated with?

25. After a concrete floor is floated and troweled, what can now be applied to give a hard, durable, and wear-resistant surface?

26. What are the cementitious materials used in applying seamless flooring?

27. What is the most common of the hardwoods used for strip flooring?

28. What cementitious material is used for exterior plastering (such as stucco)?

29. What is usually the thickness of plaster applied on masonry and wire lath?

30. In a three-coat plaster job, what is the second coat called?

31. What are gypsum plaster boards commonly known as?

32. What is gypsum plaster board with unsized paper on both sides known as?

33. Poured gypsum roofing is lightweight, fire-resistant and has the advantage of setting in what amount of time?

34. The wire lath generally used to reinforce poured gypsum roofs is of what no. gauge?

35. Which of the following U-factors, or heat loss coefficients, has the highest heat flow;

(a) 0.18

(b) 0.24

(c) 0.34

36. To find the BTU heat loss through a wall, one must multiply the U-factor by the area of the wall by what?

37. All types of roofing materials are figured by what factor?

38. In climates where there will be heavy snowfall, it is advisable to use roof pitches greater than what?

39. What is the roofing material that weighs from 700 to 810 lbs. per 100 sq. feet?

40. What are the shingles that are made from portland cement and asbestos fibers?

41. Asphalt roll roofing is used on roofs with pitches from what minimum to what maximum?

42. For what use is ternplate an excellent material?

43. Ternplate is very reactive with what metal?

44. What was probably the first ventilating window?

45. What is the wood used in the manufacture of most wood windows?

46. In the manufacture of aluminum windows, the frames, sash, and mullions are produced by what process?

47. What is the minimum tensile strength of the alloys used in aluminum windows as required by the AAMA?

48. Swing-type and revolving glass doors require what kind of glass?

49. The fire rating for a class A steel door is how many hours?

50. How are folding partitions operated?

Answers

Unit 1

1. legal description, survey
2. size, easements, lot, block, and section numbers
3. architectural survey
4. taxes, building costs, site improvement, legal fees, utilities, insurance, landscaping and planting, professional services, mortgages
5. architectural, structural, mechanical, site planning and landscaping considerations
6. yes
7. yes
8. architectural, structural, and mechanical considerations
9. architect, clerk-of-the-works, and Building Department inspectors
10. architect
11. contract drawings and specifications
12. punch lists

Unit 2

1. rock
2. rock masses detached from the ledge of which they originally formed a part
3. gravel
4. 1/4
5. decayed rock
6. hardpan
7. 80
8. 10
9. coarse dry sand, or hard clay
10. hardpan or hard shale, loose rock and hard rock
11. detached rock particles 1/4 in. to 3 in. in size, rounded and waterworn

Unit 3

1. test pit, loading platform, test borings
2. loading platform
3. load on platform times $\dfrac{B}{A+B}$
4. load of building per sq. ft.
5. 7,200 lbs.
6. 1,200 lbs.
7. 18,000 lbs.
8. 4; 24, 24
9. 3/4 in.
10. 60%
11. reduced load
12. log of test borings
13. 133, 333 lbs.

Unit 4

1. underneath the street
2. 4, 9
3. gas, electricity, telephone
4. public sewer
5. water
6. syphoned off
7. telephone and electricity
8. soil stack receives the discharge of the WC (water closet)

9. private sewage disposal system

10. public sewer or private sewage disposal system

Unit 5

1. shores
2. cap all utilities
3. removal of topsoil
4. backfill
5. the exertion of less pressure against the foundation walls
6. must be filled with concrete
7. sheet-piled and braced
8. bracing deep pits for columns
9. keeping an excavation dry during the placement of the foundation
10. clean earth

Unit 6

1. footing
2. mat, raft or boat; column or pier; wall
3. 6 in. to 8 in.
4. Douglas Fir, Southern Pine
5. precast, cast-in-place
6. mandrel
7. 8 to 10 in.
8. 20 ft.
9. concrete
10. 3 to 25
11. billet
12. cement grout
13. 5/8
14. billet
15. welded together
16. plain poured concrete, concrete block, reinforced concrete

Unit 7

1. the house drain and house sewer
2. pitched type roofs
3. cast iron
4. area of roof to be drained
5. 75 ft.
6. 150
7. trap seal
8. 1.00
9. 42
10. 67

Unit 8

1. lime, gypsum, portland cement, and masonry cements

2. dolomite limestone
3. hydrated lime
4. plasticity, workability, and water-holding capacity
5. gypsum
6. gypsum
7. Keane's cement
8. lime, silica, iron oxide, and alumina
9. standard, high early strength, modified, low heat and sulfate-resisting
10. 45 min, 10
11. 1 cu. ft., 94 lbs.
12. mortars, I and II
13. seamless flooring, terrazzo
14. asphalt concrete, paving and asphalt blocks

Unit 9

1. strength, durability
2. water-cement ratio
3. consistency, plasticity, workability
4. consistency
5. 1/5 to 1/4
6. false
7. greater than 3000 PSI.

Unit 10

1. tongue and groove boards, plywood, metal, synthetic materials
2. 1", 2"
3. 1/2" to 3/4"
4. fiberglass, steel, or special reinforced paper
5. allowing it to flow over long distances, and by excessive spading or vibrating
6. washed off
7. covering surface with wet burlap, by sprinkling periodically, by covering slabs with sand or earth that is kept moist
8. 72 hrs.
9. the loss of water
10. 300 ft.

Unit 11

1. steel rods, or wire mesh
2. 6000
3. 150,000
4. 1-way, 2-way, flat slab 2-way, flat slab 4-way
5. metal, paper
6. width 15 to 32 in., length 4 to 10 ft.
7. 2 to 3

8. 15 to 50
9. curtain walls
10. deflection

Unit 12

1. soft-mud, stiff-mud, dry-press
2. water-struck
3. sand-struck
4. 2-1/4″ × 3-3/4″ × 7-3/4″
5. 5th, 6th, or 7th
6. burning in the kiln
7. SW, MW, NW
8. false
9. face, side, cull, end, bed, and bed
10. 75%

Unit 13

1. A, B, and C
2. property specifications and proportion specifications
3. 3/8″ and 1/2″
4. A = required area in sq. ins. P = load in lbs. per sq. in., and fc = compressive strength of brick work
5. expansion joint
6. premolded, mastic, and metal
7. vertical coursing, horizontal coursing
8. weatherstruck, flush, round-rodded, V-shape
9. raked, troweled, beaded, stripped
10. 200′-0″

Unit 14

1. 75%
2. load-bearing walls, partitions and non-load-bearing partitions
3. end construction, side construction
4. LBX, LB
5. FTX, FTS
6. false
7. false
8. large sizes
9. load and non-load-bearing walls and partitions, fireproofing, back-up, foundations
10. solid, hollow
11. A, B
12. where hard, durable, easily maintained surfaces are required
13. fireproofing steel, non-load-bearing partitions

Unit 15

1. LBX, LB
2. 20′-0″
3. false
4. yes
5. FTX, FTS
6. 1/4″
7. yes
8. calculated
9. weight, color, texture, coefficient of heat transmission, strength, denseness
10. no
11. yes
12. cracking
13. dense water-resistant concrete block, surface should be treated with a waterproofing material (stucco) and painted with waterproof-type paint
14. false
15. adhesion, anchor

Unit 16

1. igneous, sedimentary, metamorphic
2. dimension (cut) stone, rubble stone (ashlar type), flagstone, monumental stone, crushed stone, stone dust
3. marble, slate, soapstone
4. granite
5. 7/8″
6. aggregate for concrete, asphaltic concrete, terrazzo, pre-cast concrete panels, artificial stone, surfacing treatment for buildings
7. 1/4″
8. dust-press, plastic
9. 6″ square
10. standard, seconds
11. nonvitreous, semivitreous, vitreous, impervious
12. trim tiles

Unit 17

1. dimension (cut) stone, rubble stone (ashlar type), flagstone
2. granite, limestone, marble
3. nonstaining, nonstaining waterproof
4. 7/8″
5. masonry back-up, concrete back-up, joining veneers, steel and ceiling
6. granite, limestone
7. granite, slate, limestone, bluestone, sandstone, soapstone

8. ceramic tile
9. ceramic mosaic
10. cement mortar, adhesive

Unit 18

1. oak, maple, birch, walnut, cherry
2. pine, spruce, fir, hemlock, redwood, cedar, cypress
3. house construction — joists, studs, rafters
4. yard lumber, structural lumber, factory or shop lumber
5. flooring, dressed and matched, lumber, siding, subflooring, sheathing
6. less than 2 in. thick and 8 in. or more in width
7. 1-1/2″ × 7-1/4″
8. plank
9. select and common
10. grade A and B suitable for natural finish, grade C and D some defects — can be covered with paint
11. classes 1, 2, 3, 4, 5
12. 5 in. or more in thickness, and 8 in. or more in width

Unit 19

1. "face," "back"
2. crossbands
3. veneer, solid wood or lumber core, particle board core
4. exterior, interior
5. 4 × 8 ft.
6. 5/16″
7. ship-lap
8. core
9. standard, tempered, low density
10. pulp, chips, fibers
11. holding power of the nails
12. insulating, acoustical

Unit 20

1. the strength of the member
2. 1″ to 2″
3. 1 ft.
4. 2-1/4″, 2-3/4″, 3″, 3-3/4″, 4″
5. 30 to 100 ft.
6. wide spans
7. does not require ties, special foundations, or buttresses
8. the steel beams
9. roofing clear spans
10. bow string truss

Unit 21

1. 4 in.
2. bridal iron (joist hangers)
3. wood bents
4. door head, jamb, sill
5. balloon, braced, western or platform, modern braced
6. western or platform

Unit 22

1. ferrous, nonferrous
2. yes
3. they corrode
4. galvanic action
5. iron, carbon
6. hard, strong, and corrosion-resistant
7. structural framing, reinforcing for concrete, forms for concrete, hollow metal work, miscellaneous and ornamental metal work
8. corrosion resistance
9. flashing, roofing, screens, tubing
10. tin
11. zinc
12. flashing, copings, roofing, bathroom accessories, hardware, galvanizing

Unit 23

1. structural, hollow metal, miscellaneous metal, ornamental metal, windows, doors, curtain walls, partitions, flashing
2. masonry
3. rolled steel sections
4. reinforced concrete slab, steel decking, metal pan, open-web joists, reinforced concrete slab on top of structural steel
5. ferrous and nonferrous
6. loose lintels, built into masonry structural shapes, stairs, railings, fences, gratings, ladders, access doors
7. ornamental metal work
8. facing type, prefabricated type, grid type, structural type
9. less thickness, lightweight, saves time in construction, adds rentable floor area, saving in heat and air conditioning losses and less air infiltration
10. concealed, exposed
11. copper, aluminum, zinc, lead, stainless steel
12. stainless steel
13. aluminum, steel

Answers

Unit 24

1. sand, soda, and lime
2. AA, A and B
3. both sides
4. rolling, sand-blasted, etched with acid, and special type roller
5. 3 to 5 times
6. by metal or glass
7. 42%
8. total radiant heat of the sun
9. 7/8″, No. 24
10. both sides
11. decorative panels and sidewalks
12. 144 sq. ft.

Unit 25

1. glazing
2. setting blocks and spacers
3. glazing compound
4. 1/4 points
5. thoroughly dried and rough projections removed
6. 40° F.
7. prime painted
8. no
9. 3/8″ expansion strips
10. lapping, attaching, condensation

Unit 26

1. no
2. no
3. 1/16″, 3/32″, 1/8″
4. 9″ × 9″, 12″ × 12″
5. will not break
6. no
7. porch roofs, screen walls, clerestory windows
8. general-purpose, for forming radii and curves, and cigarette-proof
9. glazing
10. yes
11. yes
12. tracing paper, moisture-proofing, roofing, flashing, waterproof membranes

Unit 27

1. vehicle, white paint solids (pigment)
2. liquid part
3. driers
4. the hiding power unit
5. applied at job; applied at mill, shop or factory, prime coats applied in the mill, shop, factory or on the job
6. putty, caulking compounds, wood filler paste, knot sealers, paint and varnish removers, sizing materials
7. brush, roller, spray gun
8. 60° to 80° F.
9. clean, dry, prime-coated, sized, cracks filled, rough projections removed

Unit 28

1. (a) yes, (b) yes, (c) no
2. stretching, adhesive, tape
3. yes
4. mortar base, adhesive
5. hardened, dust-proofed
6. (a) yes, (b) no, (c) yes, (d) yes, (e) no
7. no
8. heavy
9. epoxy, vinyl, urethane, acrylic, neoprene, oxychloric cement
10. gratings, plate
11. cement mortar
12. asphalt, vinyl, cork, vinyl asbestos, rubber, linoleum
13. marble, granite or limestone chips
14. divider strips
15. strip flooring, thin wood block, solid end-grain block

Unit 29

1. (a) yes, (b) no, (c) yes, (d) no, (e) yes
2. (a) cement plaster, (b) gypsum plaster, (c) Keene's cement
3. 5/8″
4. wire lath, gypsum plaster lath
5. internal and external corner beads, casings, expansion joints, strip mesh, corner mesh
6. 1-hr.
7. spackling
8. 7′-0″
9. yes
10. regular, fire-rated, backer board, decorative, insulating, vinyl coated, sheathing, laminated
11. 2, 3, or 4 sheets of plaster board laminated together

Unit 30

1. thermal insulation
2. insulation
3. blankets, batts
4. .20
5. polyethelene, aluminum foil, treated paper

6. urethane
7. low density materials
8. four
9. adhesives
10. dense

Unit 31

1. weathertight
2. waterproof
3. sun, rain, snow, sleet, ice, wind
4. roofs with pitches of 3 to 12 and greater, roofs with pitches less than 3 to 12 until they are flat
5. random, dimension, hand-split
6. design for extra load
7. adhesive tabs
8. 1 to 12 to 4 to 12
9. lapping of corrugations
10. standing seams and batten seams
11. single lock, double lock
12. plies
13. single ply, single ply with foam-type insulation, spray-on, roll-on, trowel on or bush or mop-on
14. galvanic action
15. must be easily soldered or welded
16. asbestos-cement, asphalt, slate, wood
17. yes

Unit 32

1. residential, commercial, industrial, monumental
2. double-hung
3. ponderosa pine
4. 6 to 12%
5. unglazed
6. 22,000 PSI
7. 0.062
8. swing, overhead, sliding, revolving, by-passing, folding, side-rolling
9. yes
10. yes
11. 1-3/8″, 1-3/4″
12. yes
13. 6063-T5

MID-TERM AND END-TERM REVIEW EXAMINATIONS

Units 1-8

1. conduct soil tests
2. the architect
3. owner
4. he checks all materials

5. rock
6. gravel
7. sand
8. 1 ft.
9. peat
10. 10 tons
11. 1 ton
12. building code (in area where building is to be contructed)
13. four times the design load
14. 60% of settlement under proposed load
15. 7200 lbs.
16. 1800 lbs
17. to locate the elevation of sound rock
18. bedrock
19. public utilities
20. rural areas
21. house sewer
22. soil stack
23. 7 1/2 gal. per min.
24. 6-fixture units
25. telephone and electric power
26. prevents breakage of house trap seal
27. prevents breakage of fixture trap seal
28. 12″ square, 1′-6″ deep
29. 350 ft.
30. cap or cut off gas, sewer, and water pipes
31. shores
32. inclined position
33. dry-pack
34. trench excavation
35. sheet piling
36. to keep excavation temporarily dry until foundation is installed
37. sand and gravel
38. the nature of the soil
39. 36°-53′
40. 4-6 ft. apart
41. raft or boat footing
42. cast-in-place concrete pile
43. roof area to be drained
44. 4″ per hr.
45. 15 sq. ft.
46. dolomite limestone
47. at manufacturer's plant
48. 45 min.
49. high early strength
50. coal

Units 9-16

1. strength
2. water-cement ratio

3. consistency
4. 1/4
5. 7
6. air-entraining agents
7. fill
8. permits mixes with less water
9. pine
10. steel, paper, plastic
11. excessive tamping or vibrating, correctly depositing.
12. improper mixing
13. covering surface with wet burlap, sand, straw or special papers, wetting periodically
14. temperature changes
15. loss of water
16. elastic joint filler
17. use fairly rich mix
18. 18 hrs.
19. reinforcing rods, wire mesh
20. curtain walls
21. brick
22. high quality of mechanical perfection
23. 2200 PSI
24. rate of water absorption of brick
25. natural clays
26. load-bearing
27. 3/8 in. and 1/2 in.
28. expansion joint
29. flush
30. 200 ft. long
31. 0.0000031″
32. mastic
33. soluble salts which are in the brick
34. 75%
35. fireproofing, back-up, furring, load-bearing, and non-load-bearing
36. concrete block
37. structural clay tile
38. type A
39. 20 ft.
40. 1/4″
41. cracking
42. igneous stone
43. 6,000 to 14,000 PSI
44. as a filter
45. 170 lbs./cu. ft.
46. terrazzo-type flooring
47. dark green and black
48. reddish and yellowish colors
49. gray, green, and blue
50. gray and buff

Units 17-24

1. granite
2. nonstaining

3. dampproofed, so no moisture can develop
4. 4 anchors
5. cement mortar, adhesives
6. cleavage plane
7. softwood
8. green
9. air, kiln
10. yard, structural, factory or shop
11. 1-1/2″ × 3-1/2″
12. light frame construction
13. select, common
14. dressed and matched
15. fungi
16. creosote
17. building construction
18. cabinet work, furniture, millwork
19. interior, exterior
20. 4′ × 8′
21. 3/8″
22. 5/16″
23. core
24. 5/8″ to 3″
25. 1″ to 2″
26. 24 ft.
27. 60 ft. in 1-ft. increments
28. parabolic, Tudor, Gothic
29. platform
30. balloon, braced
31. flexibility
32. 9/32″
33. galvanic
34. 2.7
35. 8.89
36. nonferrous
37. cast iron
38. structural steel framing
39. tin
40. brass
41. by gauge number
42. structural steel framing
43. exposed, concealed
44. 85%, 95%
45. sand, soda, lime
46. insulating glass
47. 0.47
48. 1 lb.
49. wire glass
50. 1.12

Units 25-32

1. 1/8″
2. painted with a prime coat
3. processed oil and white paint
4. bearing walls or partitions
5. 144 sq. ft.

6. curtain walls

7. acrylic plastic

8. 9 × 9 in. and 12 × 12 in.

9. porch roofs, screen walls, clerestory, windows

10. 0.16, 0.20

11. 10′ × 20′

12. vehicle

13. alcohol

14. 1.48

15. 60° to 80° F.

16. ratio of oil to resin

17. resistance to moisture

18. flame, pickling, sand-blasting, wire-brushing, rust remover, solvent cleaning

19. nonferrous metals

20. vapor barrier under slab and a waterproof adhesive

21. install divider strips

22. when there will be heavy wear

23. rigid

24. hardeners

25. metallic aggregates

26. oxychloric cement, epoxy, vinyl, urethane, acrylic, and neoprene

27. oak

28. portland-type cement

29. 5/8″

30. brown coat

31. sheet rock

32. backer board

33. 1/2 hr.

34. No. 12 gauge

35. 0.34

36. the temperature difference

37. the square

38. 6 to 12

39. slate

40. asbestos-cement shingles

41. 1 to 12 to 4 to 12

42. roofs

43. steel

44. casement

45. Ponderosa pine

46. extrusion

47. 22,000 PSI

48. tempered

49. 3 hrs.

50. manually or motorized

Index

Index

7931